Lecture Notes in Economics and Mathematical Systems

591

Marco Lehmann-Waffenschmidt

Economic Evolution and Equilibrium

Bridging the Gap

With 75 Figures

Springer

Professor Dr. Marco Lehmann-Waffenschmidt
Department of Economics
Economics, esp. Managerial Economics
Dresden University of Technology
01062 Dresden
Germany
lewaf@t-online.de

Library of Congress Control Number: 2007925055

ISSN 0075-8442

ISBN 978-3-540-68662-0 Springer Berlin Heidelberg New York

Springer is a part of Springer Science+Business Media

springer.com

© Springer-Verlag Berlin Heidelberg 2007

Production: LE-TEX Jelonek, Schmidt & Vöckler GbR, Leipzig
Cover-design: WMX Design GmbH, Heidelberg

SPIN 11954538   88/3180YL - 5 4 3 2 1 0   Printed on acid-free paper

Contents

Part III Economic Analysis

Part IV General Conclusions and Outlook

VIII   Contents

1

General Introduction

> It is certainly true that an actual economy will be changing all the time.
>
> J.R. Hicks

> Change is a universal phenomenon in all systems. All equilibria are temporary.
>
> K.E. Boulding

> Life is change, and without changing it would be inexplicable.
>
> N.A. Berdjajew

> Everybody knows that life is a process. But not everybody remembers that a process will be no longer a process if it reaches an equilibrium.
>
> M. Feldenkrais

> $Π\acute{α}ντα\ \overset{c}{ρ}\widetilde{ει}$.
>
> Heraklit

One of the most prominent ideas in economics undoubtedly is that of equilibrium. Even branches of economics which by their very nature are concerned with non-equilibrium states of economic systems draw on the notion of equilibrium, at least as a fundamental point of reference. Equally central to economics, however, is the idea of the evolution of an economic system over time. In fact, the understanding of an equilibrium as a final state of rest which has been borrowed from thermodynamics being prevalent in economics is obviously completely at odds with the idea of evolution. To avoid an inappropriate bipolarity of these two key concepts in economics, however, a synthesis of both seems to be desirable. Fortunately, economic theory has proposed ways to tie the two strings together. A first proposal comes from economic growth theory, which formalizes a dynamic economic system as a system of difference, or differential, equations. There equilibria mean 'equilibrium trajectories' of the whole evolution that, in a certain sense, are optimal. A particularly unsatisfactory feature of this conceptualization of an equilibrium, however, is the fact that the intertemporal optimizing approach completely predetermines the whole future of the economic system. This "closed loop" approach gives rise to the common reproach that economic theory is predominantly concerned with the question of 'how the economic system ought to behave' rather than with the question of 'how does it behave actually'. This is the point at which the new branch of evolutionary economics has made its entrance.

In contrast to growth or business cycle theory, evolutionary economics perceives the evolution of the economic system as essentially "open" to true novelties that are unforeseeable by their very nature. This view clearly makes obsolete any conception of equilibrium that resorts to the idea of a final state of rest, or to the idea of an intertemporally optimizing trajectory which is prespecified ab initio by a system of differential equations and initial conditions. To be sure, there have been attempts to reconceptualize the notion of equilibrium from the evolutionary viewpoint. However, these proposals also appear, in one way or another, to hinge on the ideas of rest. This particularly applies to the branch of nonlinear dynamics and deterministic chaotic motion. More specifically, this approach assumes the dynamic behavior of a system as being governed by a fully deterministic process, namely by iterative application of a fixed "generator" mapping. Then 'attractors' are sought, i.e. a family of states that are finally run through again and again by the system under consideration. What this approach still lacks, however, is an analytical framework for the evolving economy which allows for a new and truly 'open' conceptualization of equilibrium.

To further the latter idea we will put forth here a new attempt to synthesize the two ideas of economic equilibrium and evolution. The basic idea of our approach is to take elaborate, but equally intuitive, models of mathematical economic equilibrium theory as our starting point and to 'animate' them, or, say, 'let them evolve'.

This naturally leads us to the conception of an equilibrium as a *"transitory coordination solution"*. As we will see, this notion of equilibrium meets the requirement of an 'open evolutionary' equilibrium concept quite satisfactorily. Before we proceed to sketch our approach and our aims, we should, however, clarify our understanding of the term 'evolution' in the present study. In fact, we will adopt a broad understanding of the term 'evolution'. This means, we do not think of any connotation of progress, or directed development in any sense (*anagenesis*) when speaking of an evolving economic system. Particularly, we may, but need not necessarily, think of an evolving economic system as being governed by 'evolutionistic' rules in the sense of variation, selection, and retention. Moreover, we will employ two understandings of an evolution: a temporal understanding of an evolution as a process in historical time, and an atemporal understanding of an evolution as an "artificial" evolution generated "in the mathematical economist's laboratory". In any of these two conceptualizations an evolution consists of a succession of states of the economic system under consideration. As a general remark we would like to emphasize that throughout our whole study geometrical imagination is always a good guide for intuition.

Intuition, Scope, and Aims of the Book

Our study consists of three main Parts. In **Part I**, the concept of an evolution of economies is formalized analytically. This will be done on the basis of nine different general equilibrium models, which are henceforth refered to as the "basic models". They have been partly adopted from the literature, partly they are new. The necessary mathematical tools are introduced in the 'Mathematical Preliminaries' following this Introduction. They are mainly intuitive concepts from geometry, general topology, homotopy theory, algebraic geometry, and differential topology. At the heart of our analytical formalization of evolution lies the notion of a "continuous one-parametrization of states of the economy". This way of analytically formalizing evolutions is not only intuitive, but it also appears to be the only reasonable one for our purposes. To aid the reader's intuition, the single states of the evolution correspond to the single shots of a movie, if one compares an evolution of economies to a movie. The roots of this conceptualization as well as of further analytical treatment can be traced back to early publications by Lehmann-Waffenschmidt (1983, 1985, special aspects have been analysed by the author in 1987, 1994, 1995, 2005, 2006). Moreover, continuously one-parametrized economies have also been analyzed for instance by A. Mas-Colell in his comprehensive monograph from 1985 (Chapters 5 and 8). Indeed, both approaches have originated in complete independence of each other. The reader should note, however, that the study by Mas-Colell only provides an *analytical* treatment of one-parametrized economies, but gives no further economic applications. Nevertheless, Mas-Colell's contribution will be an important point of reference for our formal analysis in Part II of the present study. But there is a clear distinction from the mathematical viewpoint: Our constructions primarily draw on algebraic parametrized fixed point theory, whereas Mas-Colell's constructions come from the field of differential topology. It is noteworthy that our approach nowhere resorts to differentiability assumptions. All our constructions and results are solely based on assumptions of continuity.

The main task in Part I is to formalize the concept of evolutions in our nine basic set-ups and to fit these formalizations into a unifying analytical setting. This is done in order to make them accessible for an application of a crucial result from one-parametrized algebraic fixed-point theory. In Chapter 4 evolutions in three basic models from the Walrasian exchange framework, one of which is a model of large exchange economies will be formalized. In fact, this model is similar to the one used by Mas-Colell as a basis for one-parametrizations (1985, Section 5.8).

In Chapter 5, evolutions are formalized in two basic models which relax the usual assumptions of Walras' law (the budget identity) and of homogeneity of degree zero of the excess demand functions. These models are inspired by a former model by N. Schulz (1985) the purpose of which has been to model a subsystem of the system of all conceivable markets in an economy. Nevertheless, the relaxation of these two standard assumptions will prove to be of great help later in our study when a new formalization of an economy evolving in historical time is developed (Section 19.2.2).

In Chapter 6, evolutions are formalized in two models with production, tax, and subsidy schemes originally developed by T. Kehoe (1985b). Finally, in Chapter 7 evolutions are formalized in two models from the quantity constrained equilibrium framework. More precisely,a micromodel with effective demand of the Benassy type is employed, which we have slightly adapted for our purposes. Furthermore, a new model is designed with many productive sectors on a medium level of aggregation.

In **Part II** of our study, the main analytical results which will provide the basis for later applications in Part III are derived. Any proofs in Part II which employ advanced mathematical results are relegated to the appendices at the end of this monograph.

The central analytical results of this study are given in Chapter 10. Using a certain core result from parametrized algebraic fixed point theory it is shown that for any evolution of each one of the nine types introduced in Part I, there is a certain structural property of its equilibrium set. This structural property ensures the existence of what we call 'near-equilibrium paths'. This result is certainly not at all clear from the outset since even for simple examples a total indeterminacy of the equilibrium set of one-parametrizations can be observed. The intuitive geometrical meaning of a near-equilibrium path is that of a polygonal path, which lies in the graph of the Walras correspondence of the given evolution of economies. For the pure exchange framework a related result has been formerly shown by Lehmann-Waffenschmidt (1983, 1985). Another related result for a basic model of a large exchange economy has been provided by Mas-Colell (1985, 5.8.24).

A mathematical criterion is provided for checking which points lie on near-equilibrium paths. In Chapter 11, it is shown how any evolution can be approximated so that there even exists a geometrically, nicely behaved equilibrium path, i.e., a path consisting only of true equilibrium points. To our knowledge so far there is no precursor in the literature of our class of well-behaved paths and our approximating evolutions. From Mas-Colell's extension of the regular theory to the one-parametrized case merely follows the existence of approximating evolutions in the basic exchange framework. We will come back to this below. To achieve our aims, we have to accomplish three tasks. First, we must design a general class of paths that deserve the qualification "well-behaved". Second, we must provide a general construction of approximating evolutions for each of our basic models, and third we have to verify that our approximating evolutions always possess an equilibrium path from the designed well-behaved class. While this makes some analytical efforts necessary, in our eyes they are fully justified by the achievements that become possible with their help.

Actually, it is the notion of a (near-)equilibrium path which will later provide the basis for our new concept of a 'homeostatic equilibrium' of an evolution. In Section 11.1 two alternative methods of approximating evolutions of exchange economies are developed. The first one is based on piecewise linear functions, whereas the second one is based on polynomial approximation. Both methods have advantages. While the first one is completely constructive, the second one can easily be generalized to other basic frameworks.

It is noteworthy that as a byproduct of our constructions it can be shown that the graphs of the equilibrium correspondence of each of the nine basic models from Part I are "maximally well-connected". This result significantly extends the related global results on the arc-connectedness of the graph of the Walras correspondence by Y. Balasko and others (see Balasko 1988, 1996 for surveys, see also Balasko, Lang 1998 and Bonnisseau, Cayupi 1999). At this point it is also natural to examine the relationship of our results in this monograph to the results of the so-called law of demand (Hildenbrand 1989, 1994, 1998, 1999a, b). Actually, the validity of the law of demand would ensure uniqueness of the equilibrium set of any single state economy of an evolution. Then the existence of geometrically well-behaved (near)-equilibrium paths of our type would directly follow from the continuity of evolutions. Unfortunately, all theoretical and empirical results supporting the validity of the law of demand pertain to special static equilibrium model types different from any one of the nine basic models developed here. What's more, the law of demand cannot hold true for the exchange model, as simple computation shows.

Chapter 12 provides further natural interpretations and extensions of the general concept of an economic evolution developed here. Obviously our conceptualization of economic evolution by one-parametrizations gives room for two economic interpretations. On the one hand, one may emphasize the aspect that a one-parametrization connects its initial state with its terminal state. In this case, we speak of a "connection evolution". On the other hand, one may understand an evolution in this context as starting from its initial state and openly evolving in some continuous way. In this case we speak of a "course evolution". A particularly interesting question is whether for each of the basic models there is always a connection evolution for any two given economies. Fortunately, it can not only be shown that the correct answer is "yes", but also general standard constructions of connection evolutions for each basic model can be provided.

Whether one adheres to the understanding of an evolution as the performance of the economic system in historical time, or one employs the formal understanding of evolution as any succession of states, be it chronological, or artificial, it seems to be desirable to admit both cases of 'new comodities' appearing on markets and of 'old commodities' disappearing from markets during the evolution. This will be our theme in Section 12.2 where we provide the analytical constructions that are necessary to realize this in each of the nine basic models.

The structure results of the existence of (near)-equilibrium paths from Chapters 10 and 11 raise the following natural question: Is this the only structural property of the equilibrium set of evolutions that generally holds? In Chapter 13 the answer to this question will be given as affirmative for the basic models from the exchange framework in Chapters 4 and 5. Moreover, the one-parametrized extension of Mas-Colell's famous result from 1977 which extended the celebrated decomposition result of market excess demand functions by Sonnenschein, Debreu, and Mantel will be achieved. More precisely, our result shows that in the one-parametrized case of an evolution of economies there is a structural property of the equilibrium set, whereas Mas-Colell's result has verified the total lack of restriction on the equilibrium price set of a static exchange economy. As a notable corollary of our result, any two non-empty compact subsets of the price domain can be realized as the equilibrium sets of two arbitrarily close exchange economies.

Our results are also closely related to the results on the local surjectiveness of the graph of the Walras correspondence by B. Allen (1981). As we will see in Chapter 13 our results and those by B. Allen neither extend, nor contain each other, but are complementary in their characterization of the graph of the Walras correspondence. Together with the above mentioned global results by Y. Balasko and others, these results provide a fairly detailed understanding of the shape of the graph of the Walras correspondence.

In Chapter 14, we present a detailed comparison of our results with related results in the literature. As a general remark we repeat that our approach nowhere resorts to differentiability assumptions. All of our conceptualizations and results are based solely on continuity. In Section 14.1 we summarize the achievements of our results compared with the well-known global structural results on the graph of the Walras correspondence. Section 14.2 deals with the relationship of our approach and its results to the theory of regular economies and its extension to regular one-parametrizations by Mas-Colell (1985, Chapter 8). In a nutshell our conclusion is that the static regular theory produces stronger results than ours in the local sense, but if one leaves a connected component of the subspace of regular economies, these strong results break down. In this case our results have significant advantages.

There is certainly a close relationship between the theory of regular one-parametrizations and the approximation results in Section 11.1. However, there are advantages of our approach: Our method of achieving approximating evolutions by well-behaved equilibrium paths is constructive, whereas the theory of regular one-parametrizations merely provides an abstract existence result. Of course, equilibrium paths for an evolution are just selections from its equilibrium set. Thus they are non-unique in general, since the equilibrium set may well exhibit irregularities such as multifurcations, or 'thick' parts, i.e. continua. On the other hand, though they are isomorphic to linear segments,

the 'regular equilibrium paths' found by the regular theory may exhibit geometrically wild features. For instance, they may have infinite Euclidean length
as simple considerations show. Moreover, we can show here that the compart-
mentalization of the space of exchange economies as well as of the space of
exchange one-parametrizations by the subspace of critical economies and critical one-parametrizations, respectively, is fairly complicated. We emphasize
that this weakens the structure results of the regular theory considerably. Actually, slightly perturbing a critical economy, or one-parametrization, leads
to a regular economy with probability one. But the complex structure of the
subspace of critical economies makes it almost impossible to predict the properties of the obtained regular economy, or one-parametrization,respectively.
A last issue concerns the labels 'critical' and 'regular'. In fact, speaking of 'non-
regular', or 'critical', economies (one-parametrizations) means that they are
exceptional, or negligible. To be precise, this implicitly presumes a uniform
probability distribution on the space of economies (one-parametrizations).
However, so far no consistent underpinning has been provided by economic
theory which would justify the assumption of negligibility. Instead, experience
with real social systems strongly suggests that 'critical' states are not at all
negligible.

In **Part III** of the study the economic content of the preceding conceptualizations and results are explored. Following the common classification, a
distinction is made between applications on the temporal and on the atemporal field.

On the atemporal field, i.e. in the mathematical economist's laboratory,
two major strings of applications are presented (Chapter 17). The first one
has to do with the computation of equilibria, and the second one with eman-
cipating comparative statics from its paralysis through the indeterminateness
phenomenon. More precisely, it is shown that our results, in a certain sense,
achieve an extension of the well-known path following computational method
of equilibria of regular exchange economies (see e.g. Mas-Colell, 1985, Section
5.6 for a survey on the topic). The method used here works in each of our nine
basic set-ups and, particularly, is not confined to regular economies. This is,
however, at the cost of loss of algorithmic comfort.

In our second atemporal application the notorious paralysis of comparative statics caused by multiplicity of equilibria is dealt with. In fact, it is
our conviction that the multiplicity phenomenon is intrinsically linked to the
present-day way of economic thinking. Our conclusion from this is that a way
should be sought to give comparative statics a meaning, also in the multiplicity case. In Section 17.2 will be demonstrated that our preceding results
indeed provide a way to reconstruct comparative statics when equilibria are
multiple. Moreover, our main result from Chapter 13 implies that the proposed 'genetic comparative static method' in fact is *the only general way* to
achieve this.

The main economic applications of our approach and of our results, however, are on the temporal field. In Chapter 18, the methodological viewpoint

and the scope of the analysis are explained at some length. We will not strive for an analysis which is dynamic, or even evolutionary, in the strict sense, but confine ourselves to an analysis that is something like a continuous, or evolutionary comparative analysis (a *'genetic comparative analysis'*; for an evolutionary approach see e.g. Bosch 1990, Kirzner 1990, Loasby 1991, and Faber, Proops 1998, Witt 2003). From physics we have borrowed the term *'kinetic'* for our approach. As already mentioned the procedure used here is to conceptualize and formalize different types of classes of reasonable evolutions by means of continuous one-parametrizations, and then to analyze them for their general structural properties. *Kinetics* does not inquire into the causal explanation of the individual evolution of the real economic system in historical time, but searches for general regularity, or structural, properties of the dependent evolutions of the endogenous key variables. Thus, one can say, that while *dynamics* studies the *'laws of motion'* of the economic system, *kinetics* studies the *'laws of the effects of motion of the economic system'*. In this sense our temporal applications can be seen as being *complementary* to dynamics, and especially to evolutionary economics.

Having clarified our method, we will start with applications in *discrete* historical time (Chapter 19.1). The first step is to formalize evolving economies in discrete historical time for the nine basic models. This is achieved in a natural way by employing the common 'period approach'. Essentially our applications in this context are based on the atemporal applications given in Chapter 17.

The main body of our temporal applications, however, are applications in *historical time* (Sections 19.2–19.4). In Section 19.2 we begin by designing two alternative models of evolving economies in continuous time. While the first one is based on the idea of continuous flows of commodities and services, the second one provides an entirely new approach. Its main idea is to describe the evolution of a market over time by varying time intervals between two successive demand, or supply, events. In our opinion, the resulting 'frequency model' achieves a realistic theoretical framework describing an evolving economy in continuous historical time. The main ingredient of the frequency model is the basic framework of an exchange economy that relaxes Walras' law and the homogeneity assumption on excess demand functions from Section 5.1.

What are the economic achievements of the application of the analytical work from Part II to these conceptualizations of evolving economies in historical continuous time? In a nutshell, it provides the opportunity to tune equilibria, at least piecewise, continuously to their changing values when the economy undergoes an evolution. In other words, we establish the existence of a 'homeostatic equilibrium' for evolving economies. It should be emphasized that this result merely ensures the opportunity for some policy making institution to achieve a (piecewise) fine tuning of equilibrium values, but does not endogenously model the policy making institution itself. In particular, our understanding of the notion of equilibrium is not that of a description of the real state of an economy. Indeed, this is made impossible by the multiplicity of equilibria. Instead, we understand the equilibria of a given momentary state of an evolving economy here solely as momentary, or transitory, coordination solutions to this state. Consequently, it is not our concern to explain the actual states of an evolving economy, but rather to support the provision of the opportunity for 'equilibrium engineering', i.e. for continually selecting equilibrating solution values with the least possible friction. Regarding the underlying evolution of the economic system, two model approaches will be applied: In the first one the open evolution of the economic system is not touched by the "equilibrium engineering" procedure, in the second one certain "backtracking" phases in the open evolution of the economy have to be employed. We will come back to this issue shortly.

At this point, however, we would like to mention a direct application of this result to the issue of time consuming equilibrium adjustment processes. It has been known for a long time that, in general, a time consuming equilibrium adjustment process faces a moving target (e.g. Kloek 1984, for a comprehensive survey see e.g. Fisher 1983). This has already been illustrated by V. Pareto in a different context by his famous 'courbes de pursuite'. The adjustment of a moving equilibrium is symbolized by him as a running hare being tracked by a hound. To our knowledge we show for the first time that for any evolution of any of our basic models there is something like 'the path of the hare' which can be actually tracked by an agent purposed to "catch the hound" (Section 19.3).

So far the results just show that a 'frictionless equilibrium engineering', or tuning, in general is only piecewise possible, i.e., up to finitely many discrete jumps. In the final Section 19.4 we will show, however, that this deficiency can also be removed. The key idea for this is to "re-manipulate" the evolution of economies continuously without bringing new momentary states into play such that no discrete jumps in the equilibrium values are necessary when tuning them. In fact, three of the nine basic models are, from their economic conceptualization, suitable for this. These are the two models from the framework with production, taxes, and subsidies (Chapter 6) and the multi-sectoral quantity constrained model (Chapter 7). All these models have in common that they contain explicit parameters that are, in principle, accessible to an external control by some economic policy institution. These are prices and wages in the case of the quantity constrained multi-sectoral model from Chapter 7, and tax and subsidy rates in the case of the two models from Chapter 6.

In order to ensure a perfect homeostatic equilibrium, i.e., a continually frictionless tuning of equilibrium values during an evolution, it is only necessary for a policy institution to intervene at finitely many dates. In concrete terms, an effective intervention means that the evolution of the control parameters governing the evolution of the economic system is partly backtracked, i.e., is in parts repeated in a continuous way. The reader should be well aware that we have a tuning on two different levels, namely on the level of economic state parameters and on the level of equilibrium values, whereas in Section

19.2 there is only a tuning of equilibrium values. We also want to emphasize again that our result only provides the general opportunity to realize a perfect homeostatic equilibrium during an evolution to an external policy institution, but does not endogenously model policy institutions, or their actions.

The pros und cons of a continuous fine tuning of economic state parameters such as taxes, subsidies, or prices, according to the applied basic model set-up, have been, for the first time, extensively discussed in the literature during the debate on gradual versus bang-bang tax reform in the seventies (e.g. Hatta 1977, Hettich 1979). This controversy has later experienced a revival in a slightly different context, namely in the debate on macroeconomic policy design (e.g. Fellner et al. 1981, Zodrow 1985, Marangos 2002). To sum up, the following arguments are in favor of a continuous 'fine tuning' policy: Enactment of an "bang-bang", "cold turkey", or shock therapy policy entails greater administrative as well as greater social and political costs. This may largely be attributed to the agents' attitude of risk aversion and conservatism in economic affairs, which appears to be predominant in reality. Moreover, gradual control makes at least partial foresight possible for the economic agents. In economics this is generally considered as favorable for a stabilized evolution of the economy. This argument is beyond the scope of the model framework developed and employed in this book, but the reader should note that a discontinuous monitoring of equilibrium prices will not only cause sudden changes in consumed quantities, but also of individual wealth and thus of the agents' economic status. Last, but not least, tuning equilibria following a "well-behaved" path while the economy evolves is clearly much more comfortable for the agency than searching for new equilibria anywhere in the domain of all possible equilibria.

However, this does not mean that we take a one-sided position favouring a strict gradualism in economic policy making. For both positions of a gradualistic policy and a shock therapy there are striking metaphors: How would it be possible on one hand to change moving forward to moving backward other than gradually? The shock therapy position, on the other hand is favoured by the metaphor of changing from driving on the left to driving to the right in a state. We are well aware of the disadvantages of the gradualistic principle. The German reunification, for instance, may serve as an example of how political motives and uncertainties concerning the future evolution of boundary conditions may well favour a quasi bang-bang policy enactment of reforms. What we want to say is that it seems to be worthwhile investigating the conditions and opportunities for enacting a gradual, shock-free policy. A thorough assessment to decide whether a gradual adjustment, or a shock therapy policy adjustment is preferable can only be made on a case-by-case basis.

The monograph is rounded off in *Part IV* by general conclusions, an outlook on further possible research work and the Appendices *A* to *C*.

I have now reached the point where I would like to take the opportunity to thank all who have helped me with their comments and suggestions. In fact, there is a number of people who have contributed to the evolution of my personal ideas and views on my subject over the years and who have helped me to make them precise and comprehensible. Particular thanks are due to the Konrad Lorenz Institute for Evolution and Cognition Research in Altenberg near Vienna where I found the environment to do the last "finish" on this monograph, and to Barbara Feß from Springer Verlag for her help and encouragement as well as all people who gave technical support to the realization of this book.

Notations and Mathematical Preliminaries

Notations

$\mathbb{R}_+^n$, $\mathbb{R}_{++}^n$	closed (open) positive orthant of $\mathbb{R}^n$
$\mathbb{R}_-^n$, $\mathbb{R}_{--}^n$	closed (open) negative orthant of $\mathbb{R}^n$
$\partial\mathbb{R}_+^n$	boundary of $\mathbb{R}_+^n$, i.e. $\mathbb{R}_+^n\setminus\mathbb{R}_{++}^n$
Δ^{n-1}	closed $(n-1)$-dimensional unit simplex in $\mathbb{R}^n$, i.e. $\{x \in \mathbb{R}_+^n\,|\,\sum x_i = 1\}$
$\partial\Delta^{n-1}$	boundary of the $(n-1)$ unit simplex, i.e. $\{y \in \Delta^{n-1}\,|\,y_i = 0 \text{ for at least one } i = 1, \ldots, n\}$
$\mathring{\Delta}^{n-1}$	boundaryless, or open, $(n-1)$-dimensional unit simplex, i.e. $\Delta^{n-1}\setminus\partial\Delta^{n-1}$
Δ_i^{n-1}	i-facet of Δ^{n-1}, i.e. the subspace $\{x \in \Delta^{n-1}\,|\,x_i = 0\}$ of the boundary $\partial\Delta^{n-1}(i \in \{1, \ldots, n\})$
Δ_ϵ^{n-1}	for $\epsilon > 0$ the inscribed "ϵ-unit simplex", i.e. $\{x \in \Delta^{n-1}\,|\,\forall_{i=1,\ldots,n}\ x_i \geq \epsilon\}$ (note that clearly the Euclidean distance from any i-facet of Δ^{n-1} to the i-facet of Δ_ϵ^{n-1} is greater than ϵ)
$\Delta_{\epsilon_i}^{n-1}$	the i-facet of Δ_ϵ^{n-1}, i.e. the subspace $\{x \in \Delta_\epsilon^{n-1}\,|\,x_i = \epsilon\}$
$\Delta^{n-1,\alpha}$	for any real number $\alpha > 0$ the $(n-1)$-dimensional simplex $\{y \in \mathbb{R}_+^n\,|\,\sum_{i=1}^n y_i = \alpha\}$, also called the "$\alpha - (n-1)$-simplex"; thus $\Delta^{n-1,\alpha}$ is parallel to the unit simplex with intercepts α on the coordinate axes. For $\alpha \leq 1$ it is also called the "α-section of T^n" (see below); α is also called the "simplex-level" of $\Delta^{n-1,\alpha}$
$\mathring{\Delta}^{n-1,\alpha}$	open $\alpha - (n-1)$-simplex $\Delta^{n-1,\alpha} \cap \mathbb{R}_{++}^n$
$\langle v_1, \ldots, v_{k+1}\rangle$	for $k + 1$ points $v_1, \ldots, v_{k+1}$ in $\mathbb{R}^n$ the m-dimensional simplex generated by them, i.e. their convex hull

	$(m \leq k \leq n)$
S^{n-1}	$(n-1)$-dimensional unit sphere, i.e. $\{x \in \mathbb{R}^n \mid \|x\| = 1\}$
S^{n-1}_+	closed positive part of S^{n-1}, i.e. $\{x \in \mathbb{R}^n_+ \mid \|x\| = 1\}$
S^{n-1}_{++}	strictly positive part of S^{n-1}, i.e. $\{x \in \mathbb{R}^n_{++} \mid \|x\| = 1\}$
S^{n-1}_ϵ	for $\epsilon > 0$ the contained closed ϵ-unit sphere, i.e. $\{x \in S^{n-1}_+ \mid \forall_{i=1,\ldots,n}\ x_i \geq \epsilon\}$
$T^{n,\alpha}$	for any real number $\alpha > 0$ the orthogonal projection of $\Delta^{n,\alpha} \subset \mathbb{R}^{n+1}_+ := \mathbb{R}^n_+ \times \mathbb{R}_+$ into the coordinate hyperplane $\mathbb{R}^n_+$, i.e. $T^{n,\alpha} := \{x \in \mathbb{R}^n_+ \mid \sum_{i=1}^n x_i \leq \alpha\}$, also called "embedded α-n-simplex"
$\overset{\circ}{T}{}^{n,\alpha}$	the open embedded α-n-simplex $\{x \in \mathbb{R}^n_{++} \mid \sum_{i=1}^n x_i < \alpha\}$
T^n	abbreviation of $T^{n,1}$, also called "embedded n-dimensional unit simplex"
0^n	the null vector of $\mathbb{R}^n$
T^n_0	the pointed embedded n-dimensional unit simplex $T^n \backslash \{0^n\}$
$T^{n,\alpha}_\gamma$	for arbitrarily large real positive α and arbitrarily small real positive γ the "inscribed embedded α-γ-n-simplex"

$$\{x \in \mathbb{R}^n_+ \mid \sum_{i=1}^n x_i \leq \alpha \text{ and } x_i \geq \gamma \text{ for all } i = 1, \ldots, n\}$$

T^n_γ	$T^{n,1}_\gamma$
$\{pt\}$	the single point space (singleton set)
e_n	the vector $(1, 1, \ldots, 1)$ of $\mathbb{R}^n$
e^i	the i-th unit vector $(0, \ldots, 0, 1, 0, \ldots, 0)$ of $\mathbb{R}^n$
$(x_1, \ldots, \hat{x}_i, \ldots, x_n)$	the $n-1$-vector $(x_1, \ldots, x_{i-1}, x_{i+1}, \ldots, x_n)$
$x \geq y,\ x > y$	for n-vectors x and y means that the weak (strong) inequality holds for every component
$\overline{xy}$	for n-vectors x and y the (straight line) segment with endpoints x and y
x'	for a column vector $x \in \mathbb{R}^n$ the transposed row vector
$Y \backslash X$	for spaces $X \subset Y$ the difference set $\{y \in Y \mid y \notin X\}$
X^c	for spaces $X \subset Y$ the complement of X in Y, i.e. $Y \backslash X$
coX	for a subspace $X \subset \mathbb{R}^n$ denotes the convex hull of X in $\mathbb{R}^n$

$dist(F, G)$ means for any two nonempty compact subsets F, G of a metric space X the Hausdorff distance, i.e.

$$\min\{\epsilon \geq 0 | F \subset B_\epsilon(G) \text{ and } G \subset B_\epsilon(F)\}$$

where $B_\epsilon(Y) = \{x \in X | d(x, y) < \epsilon \text{ for some } y \in Y\}$ for any $Y \subset X$

$B_r^n(x)$ the closed n-ball with center x and radius $r \geq 0$, i.e.
$\{y \in \mathbb{R}^n | \, ||y - x|| \leq r\}$

$im \, f$ for a mapping $f : X \to Y$ the image $f(X) \subset Y$

$Fix \, f$ for a self-mapping $f : X \to X$ the fixed point set, i.e. $\{x \in X | f(x) = x\}$

$Fix \, F$ for a homotopy $F : X \times [0, 1] \longrightarrow X$ the set $\{(x, s) \in X \times [0, 1] | F(x, s) = x\}$

$X \times [0, 1]$ homotopy space, i.e. the domain of a homotopy $F : X \times [0, 1] \to Y$; due to their geometrical shape the special homotopy spaces $\Delta^{n-1} \times [0, 1]$ and $T^{n-1} \times [0, 1]$ are called "homotopy prisms"

$X \times \{s\}$ for $0 \leq s \leq 1$ the "s-slice" of the homotopy space $X \times [0, 1]$

$\mathbb{N}$ natural numbers including 0

$M(n \times m; \mathbb{R})$ the set of $n \times m$-matrices with real entries

$|A|$ for an $m \times n$-matrix $A = (a_{ij})$ with real entries the $m \times n$-matrix $(|a_{ij}|)$ of absolute values of the entries

Mathematical Preliminaries

Now we are going to provide the reader with the formal standard notions from general and algebraic topology and algebraic geometry as well which will play an important role in our analysis[1]. Our exposition will be self-contained as regards our subsequent analysis. The reader who still misses further background informations is referred to the relevant textbook literature.

At the heart of our formalizations stands the notion of a continuous one-parametrization. Generally a *continuous one-parametrization,* or *evolution, homotopy, deformation, family* or *perturbation,* is a continuous mapping $F : X \times [0, 1] \longrightarrow Y$ where X and Y are topological spaces. To be sure, the notion of a continuous one-parametrization has intuitive appeal since it can be viewed as a continuous one-parameter family of "ordinary" continuous mappings $(F_s)_{s \in [0,1]} : X \longrightarrow Y$ where $F_s(x) := F(x, s)$. For instance, any continuous movement process is an example of a continuous one-parametrization (cf. Figure 2.1). The subspace $X \times \{s\}$, $s \in [0, 1]$, is called the s-*slice*

---

[1] As this Section consists of a collection of definitions we will omit the term "definition" throughout.

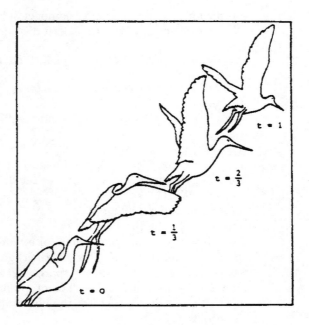

Fig. 2.1: Formal Representation of a Movie as a Homotopy

of the *homotopy space* $X \times [0,1]$ (corresponds to the snap–shot photo at time $t = s$ in Figure 2.1). We furthermore call s the *homotopy, deformation,* or *evolution, parameter* and the mapping $F_s(-) = F(-,s)$ the *s-state mapping of the one-parametrization* F. Thus one can visualize a one-parametrization $F : X \times [0,1] \longrightarrow Y$ for Euclidean subspaces $X \subset \mathbb{R}^m$ and $Y \subset \mathbb{R}^n$ by the continuous evolution of the graphs of the n component functions $F_i(x,s)$. *Evolutions of economic systems* which we will employ in our study will always be formally representable by one-parametrizations. Recall from the General Introduction that in our study we neither restrict the term 'evolution' to economic systems which are characterized by 'evolutionary' (technical) progress, nor do we even stick to the narrow understanding of evolutions as necessarily being evolutions over (historical) time. Rather, we will introduce evolutions of economic systems in the general notion of *any* continuous changes governed by a scalar parameter s. In Part III of our study we will study evolutions of economies in both interpretations of the evolution parameter s: in the technical atemporal interpretation, and in the interpretation as elapsing historical time.

Clearly, one can *combine,* or say *compose,* two homotopies $F^1, F^2 : X \times [0,1] \longrightarrow Y$ when $F_1^1 = F_0^2$, i.e. $F^1(x,1) = F^2(x,0)$ for all $x \in X$. We also say that the obtained homotopy $F : X \times [0,1] \longrightarrow Y$,

$$F(x,s) = \begin{cases} F^1(x,2s) & \text{if } s \in [0,1/2], \\ F^2(x,2s-1) & \text{if } s \in [1/2,1], \end{cases}$$

is the *composition* of the two homotopies F^1 and F^2, or the *composite homotopy* of F^1 and F^2. Composing k homotopies in this way accordingly leads to a $(k-1)$-*fold composite homotopy*.

A *contractible* topological space X is *homotopic* to the single point space, i.e. there is a homotopy $F : X \times [0,1] \longrightarrow X$ with $F(-,0) = id_X$, and $F(-,1)$ is a constant mapping into some point $x \in X$. Contractible spaces are special examples of *acyclic* spaces. The *Lefschetz number* of a space is an algebraic topological characteristic. For an acyclical space it is $+1$ (the interested reader is referred to Brown, 1971, II). A subspace X of $\mathbb{R}^n$ which is not convex can still be *star-shaped*, that means there is a point $x_0 \in X$ such that any two points $x,y \in X$ can be connected by the two segments $\overline{xx_0}$ and $\overline{x_0y}$. Clearly, a star-shaped space is contractible.

If the homotopy space equals the unit interval a special type of a homotopy called *"path"* obtains. Indeed, the concept of a *Euclidean path* $w : [0,1] \longrightarrow \mathbb{R}^n$ will be crucial for our study, and it especially gives rise to the following concept of a connected component of a space.

A *connected component* Z of some topological space X cannot be separated into two disjoint open subsets, i.e. there are no disjoint open subsets A, B of X with $(A \cup B) \cap Z = Z$. For any two points x,y of a *path (connected) component* Z' of X there is a continuous path $w : [0,1] \longrightarrow X$ with $w(0) = x$, $w(1) = y$. w *is a path in X connecting x with y.* A path connected component is maximal with this property.

One has to distinguish carefully between the notion of a path w and of its *arc*, i.e. its image $w[0,1]$ in $X \subset \mathbb{R}^n$. Identifying $[0,1]$ with $\{y\} \times [0,1]$ a path $w : [0,1] \longrightarrow X$ can also be viewed as a continuous one-parametrization $w : \{y\} \times [0,1] \longrightarrow X$ of its arc $w[0,1]$. Note particularly that in a graphic representation the parameter $t \in [0,1]$ in general is *not identifiable on the coordinate axes*. The *Euclidean length of a path* $w : [0,1] \longrightarrow X \subset \mathbb{R}^n$ is defined as $\sup_{W_k} L(w,W_k)$ where W_k denotes a subdivision of $[0,1]$ by $k+1$ points $0 = t_0 < t_1 < \ldots < t_k = 1$ and $L(w,W_k) := \sum_{j=1}^{k} d(w(t_{j-1}), w(t_j)) = \sum_{j=1}^{k} ||w(t_j) - w(t_{j-1})||$. If $\sup_{W_k} L(w,W_k)$ is finite then one says that w is of finite length, or w is *rectifiable*. It is well-known that a path w is rectifiable if and only if each of its component functions w_i, $i = 1, \ldots, n$, is of *bounded variation over* $[0,1]$, that means

$$\sup_{W_k} \sum_{j=1}^{k} ||w_i(t_j) - w_i(t_{j-1})|| < \infty.$$

The everyday connotation of the term 'path' clearly is 'to be viable, or passable' in the intuitive geometrical sense. This is also our intuition in this study. Unfortunately, arcs of *continuous* paths can still have wild shapes as the following examples show: the graph of the continuous function $x \cdot \sin 1/x$ on

$[-1, 1]$ has *infinite length* (one estimates from below by the divergent harmonic series). But even if the arc of a continuous path is of finite length, it still may *oscillate*, or *tremble*, infinitely often, as the function

$$x \mapsto \begin{cases} x^3 \sin 1/x, & x \in [-1, 0[\ \cup \ ]0, +1] \\ 0, & x = 0 \end{cases}$$

shows ('a damped oscillation', see Fig. 5).

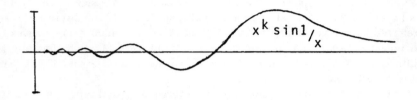

Fig. 2.2: Damped Oscillation

Fortunately, there is a way to analytically design a broad class of paths whose arcs are really "nice" in the intuitive geometrical sense. In other words they do not display any features of impassableness. We will come back to this later in our study (Section 11.1).

A *gluing (hat) function* $\alpha : \Delta^{n-1} \longrightarrow [0, 1]$ which continuously glues some continuous function $f : \Delta^{n-1} \longrightarrow \mathbb{R}^m$ with some other continuous function $g : \Delta^{n-1} \longrightarrow \mathbb{R}^m$ over the area $\Delta^{n-1}_\epsilon \backslash \mathring{\Delta}^{n-1}_{2\epsilon}$ such that f prevails on the inner part $\Delta^{n-1}_{2\epsilon}$ and g on the boundary area $\Delta^{n-1} \backslash \Delta^{n-1}_\epsilon$ is a continuous function with the properties

$$\alpha\big|_{\Delta^{n-1} \backslash \Delta^{n-1}_\epsilon} \equiv 0$$
$$\alpha\big|_{\Delta^{n-1}_{2\epsilon}} \equiv 1.$$

The *glued function* is given by the *convex*, or *linear*, *combination*

$$\alpha(x)f(x) + (1 - \alpha(x))g(x).$$

An "*(affine) simplex* of dimension $k \geq 0$ embedded into $\mathbb{R}^n$" is the convex hull of $k + 1$ different points $v_0, v_1, \ldots, v_k$ in $\mathbb{R}^n$ *which moreover are in general position*. The latter means that the affine linear subspace of $\mathbb{R}^n$

$$\left\{ y \in \mathbb{R}^n \ \middle|\ y = v_0 + \sum_{i=1}^{k} \lambda_i(v_i - v_0), \ \lambda_i \in \mathbb{R} \right\}$$

spanned by $v_0, v_1, \ldots, v_k$ is *not* spanned by any subset of $\{v_0, v_1, \ldots, v_k\}$. Thus, the *affine simplex* $\langle v_0, \ldots, v_k \rangle \subset \mathbb{R}^n$ *generated by* $v_0, v_1, \ldots, v_k$ is the subspace

$$\left\{ \sum_{i=0}^{k} \lambda_i v_i \,\middle|\, \sum_{i=0}^{k} \lambda_i = 1, \quad \text{every} \quad \lambda_i \geq 0 \right\}.$$

$v_0, \ldots, v_k$ are also called the *vertices* of the simplex $\langle v_0, \ldots, v_k \rangle$. Each subset $\{v_{i_0}, \ldots, v_{i_l}\}$ of the set of vertices spans an affine subsimplex $\langle v_{i_0}, \ldots, v_{i_l} \rangle$ of $\langle v_0, \ldots, v_k \rangle$. $\langle v_{i_0}, \ldots, v_{i_l} \rangle$ is also called an *l*-dimensional *face* of the simplex $\langle v_0, \ldots, v_k \rangle$. In this terminology, the vertices are precisely the 0-faces, and the 1-faces are called the *edges* of the simplex. The maximal dimension of an affine simplex in $\mathbb{R}^n$ is clearly n.

A subspace X of $\mathbb{R}^n$ is a *finite simplicial complex* $\sum_X$ *of dimension k* if it is the union of finitely many affine simplices of dimension $\leq k$ which satisfy the following rules of adjacency: for any simplex from $\sum_X$ each of its faces also belongs to $\sum_X$. The intersection of any two simplices from $\sum_X$ is either empty or is a common face.

One also calls $X = \bigcup_{\sigma \in \sum_X} \sigma$ the *support* of the simplicial complex $\sum_X$, and says that X is *finitely simplicially decomposed*, or *finitely triangulated*, by the simplicial complex $\sum_X$. In this study we will only deal with finite simplicial decompositions of simple Euclidean subspaces like Δ^{n-1} or Δ_ϵ^{n-1}, for instance.

There is obviously no difficulty to *extend* a given simplicial triangulation $\sum_X$ of Δ_ϵ^{n-1} to Δ^{n-1}, i.e. to provide a simplicial decomposition $\sum_X'$ of Δ^{n-1} whose restriction to Δ_ϵ^{n-1} equals $\sum_X$. Furthermore it is straightforward for these simple spaces to obtain a finite triangulation $\sum_X$ for any two given triangulations $\sum_X'$ and $\sum_X''$ which is a *common refinement* of $\sum_X'$ and $\sum_X''$, i.e. which contains both complexes $\sum_X'$ and $\sum_X''$ as subcomplexes.

The spaces Δ^{n-1} and S_+^{n-1} are standard examples of *neighborhood retracts in a Euclidean space*. Generally, a *Euclidean neighborhood retract A* in $\mathbb{R}^n$ is a subspace which is a *retract* of some of its neighborhoods, i.e. there is a neighborhood $\mathcal{U}(A)$ of A in $\mathbb{R}^n$ and a continuous mapping

$$r : \mathcal{U}(A) \longrightarrow A \text{ with } r|_A = id_A.$$

For the purpose of our present study, i.e. for the equilibrium analysis of evolutions of economic systems, the notion of a homotopy is still not quite satisfactory. The reason for this is that the continuity of a homotopy is a fairly weak property still allowing for some pathologies of the one-parametrized family of state mappings if the domain is not a compact space. More formally, the continuity of a homotopy F is equivalent to C^0-uniform convergence of the state mappings F_s on compacta, i.e. to convergence of the state mappings F_s on any compact subset $A \subset X$ with respect to the maximum norm. However, if X is an *open* subspace of $\mathbb{R}^n$ this admits for instance the following pathology for a continuously one-parametrized family of excess demand functions $(\zeta_s)_{s \in [0,1]} : \mathring{\Delta}^{n-1} \longrightarrow \mathbb{R}$ (see Figure 2.3): The sequence of excess demand functions $(\zeta^k)_{k=1,2,\ldots}$ obviously converges to the excess demand function ζ^0

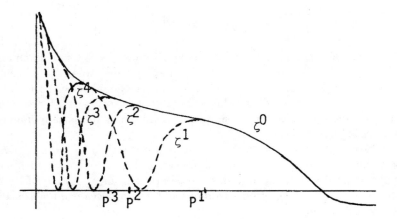

Fig. 2.3: Convergence Pathology

on compacta, though there is an increasing deviation of the functional values for the critical arguments $P^1, P^2, \ldots$. This is possible since the critical arguments run to the boundary of the non-closed domain. Clearly such a behavior strongly contradicts the intuition underlying the notion of a continuous evolution of economic behavior functions. That means that neighboring state functions of an evolution should have similar values *on their whole domain* – not only on compacta. Thus, throughout our whole study we will employ the stronger concept of *overall C^0-uniform convergence* for one-parametrizations instead of mere continuity, i.e. the state functions must converge on their whole domain with respect to the usual supremum norm.

Given subspaces $X \subseteq Y \subseteq \mathbb{R}^n$, a real $\epsilon > 0$, and a function $g : X \longrightarrow \mathbb{R}^m$, we say that a function $f : Y \longrightarrow \mathbb{R}^m$ *ϵ-approximates g uniformly on X* when the restriction f^3_X is in the ϵ-neighbourhood of g, i.e.

$$\|f(x) - g(x)\| = \sqrt{\sum_{i=1}^{m}(f_i(x) - g_i(x))^2} \ < \ \epsilon \quad \text{for all } x \in X.$$

Now we are going to introduce the concept of *semi-algebraic subsets* of $\mathbb{R}^n$. We will employ these sets since they have very nice geometrical properties and help us to formalize the notion of "nice paths". Let us first recall some elementary definitions from algebraic geometry: a *polynomial in n-variables over $\mathbb{R}$* is a continuous mapping $f : \mathbb{R}^n \longrightarrow \mathbb{R}$ of the form

$$f(x_1, \ldots, x_n) = \sum a_{i_1 \ldots i_n} \cdot x^{i_1} \cdot \ldots \cdot x^{i_n}$$

where the coefficients $a_{i_1 \ldots i_n}$ are fixed real numbers and the sum is taken over a *finite* set of n-tuples $(i_1, \ldots, i_n)$ of positive integers. $\mathbb{R}[x_1, \ldots, x_n]$ denotes the set of all polynomials in n variables over $\mathbb{R}$. Thus, $\mathbb{R}[x_1, \ldots, x_n]$ particularly contains all linear equations with real coefficients in n variables.

For our purposes the *zero sets of polynomials* are crucial. A subset $A \subset \mathbb{R}^n$ is called *algebraic* if it is the simultaneous zero set of a finite number of polynomials $f_1, \ldots, f_r \in \mathbb{R}[x_1, \ldots, x_n]$, i.e.

$$A = \{x \in \mathbb{R}^n | f_1(x) = \ldots = f_r(x) = 0\}.$$

A subset $A \subset \mathbb{R}^n$ is called *semi-algebraic of the first kind,* if there exists a polynomial $f \in \mathbb{R}[x_1, \ldots, x_n]$, such that

$$A = \{x \in \mathbb{R}^n | f(x) > 0\} = f^{-1}(]0, \infty[).$$

A subset $X \subset \mathbb{R}^n$ is called *semi-algebraic,* if it can be written in the form

$$X = \bigcup_{i \in I} \bigcap_{j \in J} (A_{ij} \setminus B_{ij})$$

where I, J are arbitrary finite index sets, and the sets A_{ij}, B_{ij} are semi-algebraic of the first kind. Evidently, the semi-algebraic subsets of $\mathbb{R}^n$ form a rich class.

Notice the following obvious *properties of semi-algebraic sets* which will be helpful for our later analysis:

(i) If $X \subset \mathbb{R}^n$ is algebraic, then X is semi-algebraic.
(ii) Let $f_1, \ldots, f_r \in \mathbb{R}[x_1, \ldots, x_n]$ be polynomials. Then the set

$$X = \{x \in \mathbb{R}^n | f_1(x) > 0, \ldots, f_r(x) > 0\}$$
$$= \bigcap_{i=1}^{r} f_i^{-1}(]0, \infty[)\}$$

is semi-algebraic.
(iii) If the sets $X, Y \subset \mathbb{R}^n$ are semi-algebraic, then the sets $X \cap Y$, $X \cup Y$, $X \setminus Y$ are also semi-algebraic.

As our last geometrical concept we introduce the central projection mapping in $\mathbb{R}^n$ together with its inverse. The *central projection from the origin* is the homeomorphism

$$\varphi : T^n \setminus \Delta^{n-1} \xrightarrow{\approx} \mathbb{R}^n_+$$

$$x \mapsto \left(\frac{1}{1 - \sum\limits_{i=1}^{n} x_i} \right) x.$$

That means φ is one-to-one and onto, and φ and its inverse φ^{-1} are both continuous $(\varphi^{-1}(y) = \left[1 - \frac{\sum_{i=1}^{n} y_i}{1 + \sum_{i=1}^{n} y_i} \right] y)$. By φ any vector $x \in T^n \setminus \Delta^{n-1}$

is *radially* stretched by the factor $\dfrac{1}{(1 - \sum_{i=1}^{n} x_i)} \geq 1$. This factor grows beyond all finite bounds when x approaches Δ^{n-1}. For a point $x \in \mathbb{R}_+^n$ we denote by $\sum_{i=1}^{n} x_i$ the 'simplex level' of x. Thus, the simplex level of the image point $\varphi(x)$ is $\dfrac{\sum_{i=1}^{n} x_i}{1 - \sum_{i=1}^{n} x_i}$. φ particularly is "simplex-preserving", i.e. it maps an α-section $\Delta^{n-1,\alpha}$, $0 < \alpha < 1$, of $T^n \backslash \Delta^{n-1}$ one-to-one onto $\Delta^{n-1,\frac{\alpha}{1-\alpha}}$, or in other words, the simplex-level of the image simplex equals $\frac{\alpha}{1-\alpha}$. For example, the $(n-1)$-dimensional unit simplex is the φ-image of $\Delta^{n-1,\frac{1}{2}}$ (see Figure 2.4).

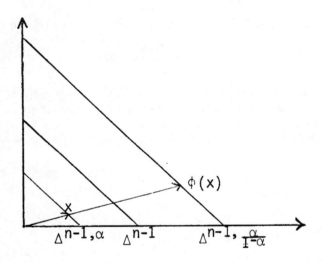

Fig. 2.4: Variants of the (n-1)-Dimensional Unit Simplex

A sequence $(\mu_n)_{n \in \mathbb{N}}$ of measures on a metric space T is said to *converge weakly* to a measure μ on T if $\int f d\mu_n \xrightarrow{n \longrightarrow \infty} \int f d\mu$ for any continuous and bounded real-valued function f on T. (For further details see Hildenbrand (1974), Section I. D, or Mas-Colell (1985), Chapter 1.E, and the references given there. A compilation of equivalent characterizations of weak convergence is given by Lehmann-Waffenschmidt (1985), pp. 56–57.)

Finally, let us summarize the notion of an *explicit finite exchange economy* in the notations used by Mas-Colell (1977) (cf. also Hildenbrand, 1974, and Shafer/Sonnenschein, 1982). By an explicit finite exchange economy we denote an exchange economy with finitely many commodities and finitely many agents characterized by preferences and initial endowments. Formally, an *explicit finite exchange economy* with l agents and l commodities is denoted by $(\precsim_i, \omega_i)_{i=1}^l$. $\omega_i \in \mathbb{R}_{++}^l$ is agent i's initial endowment vector. His preference relation $\precsim_i$ on the commodity space $\mathbb{R}_+^l$ is an element of $\mathcal{P}_{mo}^0$, the space of

the continuous, monotone, and strictly convex preference relations on $\mathbb{R}_+^l$. Thus, *agent i's individual demand set* for any given price vector $p \in S_{++}^{l-1}$, i.e.
$$g_i(p) = \{x \in \mathbb{R}_+^l \mid px \leq p\omega_i \text{ and } \forall_{y \in \mathbb{R}_+^l} \, py \leq p\omega_i \text{ implies } y \precsim_i x\},$$ is a singleton. Given a continuous and monotone preference relation $\precsim$ on $\mathbb{R}^l$ the *closed upper contour set*, or *indifference-or-preference set*, of any point $x \in \mathbb{R}^l$ is denoted by $\psi(x) := \{y \in \mathbb{R}^l \mid x \precsim y\}$.

Given an *l*-tuple of pairs $(\precsim_i, \omega_i)_{i=1}^l$ with $\precsim_i \in \mathcal{P}_{mo}^0{}_{sco}$ and $\omega_i \in \mathbb{R}_{++}^l$ for all $i, \ldots, l$ the function

$$f_i : S_{++}^{l-1} \longrightarrow \mathbb{R}^l$$

$$p \mapsto \{x \in \mathbb{R}_+^l \mid px \leq p\omega_i, \text{ and } py \leq p\omega_i \text{ for any } y \in \mathbb{R}_+^l \text{ implies } y \precsim_i x\} - \omega_i =$$

$$g_i(p) - \omega_i$$

is the *individual excess demand function derived from* $(\precsim_i, \omega_i)$, and

$$f(p) := \sum_{i=1}^l f_i(p)$$

is the *market excess demand function derived from the l-consumer exchange economy* $(\precsim_i, \omega_i)_{i=1}^l$ (cf. Mas-Colell, 1977, p. 118). The function f has the usual properties: it is continuous and bounded from below, i.e. there is some $k < 0$ such that $f(p) > ke$, and satisfies Walras' law and the desirability condition: for every sequence (p^n) in S_{++}^{l-1} which converges to a point of the boundary ∂S_+^{l-1} one has $\|f(p^n)\| \longrightarrow +\infty$. One also says that the pair $(\precsim_i, \omega_i)$ (the economy $(\precsim_i, \omega_i)_{i=1}^l$) *generates* the individual (the market) excess demand function $f_i (f)$ on S_{++}^{l-1}.

Naturally the notion of a *continuous one-parametrization of exchange economies with l agents* is central for our analysis. Formally it is given by a continuous one-parametrization

$$(E_s)_{s \in [0,1]} : \{1, \ldots, l\} \times [0,1] \longrightarrow \mathcal{P}_{mo}^0{}_{sco} \times \mathbb{R}_{++}^l$$
$$(i, s) \mapsto (\precsim_{i_s}, \omega_{i_s}).$$

One also could say that $(E_s)_{s \in [0,1]}$ is an *l*-tuple of continuous paths $(\precsim_{i_s}, \omega_{i_s}), i = 1, \ldots, l$, in the space $\mathcal{P}_{mo}^0{}_{sco} \times \mathbb{R}_{++}^l$. While the meaning of a continuous path $(\omega_{i_s})_{s \in [0,1]}$ in $\mathbb{R}_{++}^l$ is clear, we still have to explain what to understand under a continuous path of preferences $(\precsim_{i_s})_{s \in [0,1]}$ in $\mathcal{P}_{mo}^0{}_{sco}$. Here we will employ the following intuitive notion which is in the lines of Debreu (1969) (cf. also the discussion by W. Hildenbrand (1974), Notes 1.2, pp. 108–109). Roughly speaking a continuous path $(\precsim_{i_s})_{s \in [0,1]}$ in $\mathcal{P}_{mo}^0{}_{sco}$ means that the upper contour sets change continuously in the Hausdorff sense. More formally: the closed upper contour set $\psi_{i_s}(x)$ of any $x \in \mathbb{R}_+^l$ with respect to some fixed preference relation $\precsim_{i_s}$ is Hausdorff-continuously deformed when $s \in$

$[0,1]$ varies continuously, i.e. $\psi_{i_s}(x)$ is Hausdorff-continuously deformed into upper contour sets $\psi_{i_s}(x_s)$ with suitable vectors $x_s \in \mathbb{R}_+^l$. Particularly this means that any upper contour set $\psi_{i_s}(x^0)$ for fixed $x^0 \in \mathbb{R}_+^l$ varies Hausdorff-continuously with the continuously varying homotopy parameter $s \in [0,1]$.

Conceptualization and Definition of Evolutions
of Economies in Four General Equilibrium
Frameworks

3

Introduction to Part I

In Part I nine static equilibrium models will be introduced. They provide the basic set–ups in which the notion of an evolution will be defined. Some of our *basic set–ups* are adopted from the literature (Chapters 4 and 6), whereas the others are new (Chapters 5 and 7). Naturally, these are also inspired by existing frameworks.

These nine basic models are grouped into four equilibrium frameworks: the Walrasian exchange framework (Chapter 4), an exchange framework which relaxes the traditional assumptions of Walras' law and homogeneity of degree zero of the excess demand functions (Chapter 5), the framework with production, taxes, and subsidies developed by T. Kehoe (1982, 1985 a,b, Chapter 6), and the quantity constrained temporary fixed price framework (Chapter 7).

The notion of an evolution of an economic system usually has the connotation of historical time. In the present study, we take the view of a theorist who is generally interested in a comprehensive equilibrium analysis of evolving economies – be it evolutions over historical time, or evolutions in logical, artificial time in the laboratory. Thus, in this study, an "evolution of economies" will denote any succession of states of the economic system under consideration, whether the states are changing, or not. This can be compared with a cinematic study of a movement process in sports. On the one hand, one can use it for recording and representing the movement process in real time. On the other hand, if one wants purposed to perform specific analyses, one can employ certain cinematographic techniques that make it possible to re-manipulate the representation by slow motion, fast motion, backtracking, and so on.

The metaphor of a movie recording naturally suggests how to analytically formalize an evolution. In fact, this is most naturally achieved by a one–parametrization, or say a one-parameter family, of single shot states of the economic system considered. This will be our approach in this study. In order to stress our generalized usage of the notion of an evolution with respect to the aspect of historical time, we generally use the symbol s, and not t, for the scalar evolution (or say, variation, or deformation) parameter throughout the book. In his book, Mas-Colell (1985, Section 5.8) introduces a generalized concept of parametrizing economies by parameters which are not necessarily scalars. Nevertheless, he attributes the greatest importance to the one-para-metrized case (ibid., p. 235). The analysis of one-parametrized economies by Mas-Colell is, however, confined to a certain of large exchange economies (see Chapter 7 below).

Each chapter of Part I is devoted to one model framework and will be organized according to the same scheme: Each section deals with one specific basic model and is further divided into *three subsections*. Subsection (i) motivates the model and specifies the notions of an economy, an equilibrium, and an evolution of economies in this set–up. Subsection (ii) presents the construction of the most important technical tool of our analysis, a continuous self–mapping of some compact Euclidean space which equivalently transforms the zero–problem of the existence of equilibria into a fixed–point problem. This is done in order to make the equilibrium analysis amenable to the powerful tools of the *one–parametrized fixed–point theory* (see Part II). We will call the addressed mapping an 'equilibrium equivalent self–mapping'. Finally, subsection (iii) contains the verification that any admissible evolution of economies in the present basic set–up actually yields a *continuous* one-parametrization of associated equilibrium equivalent self–mappings. In fact, it is this property of an evolution of economies as we formalize it that will turn out to be the essential prerequisite the results in Part II. Fortunately, the technical work done in the subsections (ii) and (iii) of the chapters in Part I settles the major part of the technical efforts that are necessary for our central analytical results in Chapters 10 and 11 in Part II.

Evolutions in the Traditional Walrasian Exchange Equilibrium Framework

We start our analysis in the traditional Walrasian general equilibrium framework of pure exchange. There are two main reasons for us to do so. The first reason is that, as most economists certainly will agree, the Walrasian general equilibrium framework stands at the very heart of economics as a fundamental point of reference. Or, as Balasko (1988, p. viii) puts it: "... for them [the pure exchange economies], the hidden and intricate structure of the equilibrium model is most easily brought to light. Besides the intrinsic interest in such an undertaking, the insights gained from understanding the mathematical structure of the simpler pure exchange model can be invaluable when dealing with more general models." Thus, it will hardly be surprising that also for our present analysis the Walrasian framework will turn out to be most useful from the viewpoint of economic intuition as well as from the viewpoint of technical convenience. Indeed, several times in our study we easily can transfer an analysis which we have carried out for the Walrasian set–up to other basic set–ups.

The second reason mentioned above derives from the well-known result of indeterminateness of the exchange framework which was developed in the early seventies by Sonnenschein, perfected by Debreu (1974) and further generalized by Mas-Colell (1977) (see also Shafer/Sonnenschein (1982) for a survey). Our main concern is with the result by Mas-Colell (1977) who has shown that the feature of indeterminateness also pertains to the equilibrium set itself. More precisely, this means that *any compact set* of the price simplex can be realized as the equilibrium set of some reasonable exchange economy described by individual preferences and endowments.

In Part II of our study we will extend the examination of indeterminateness from the static level to the one-parametrized level. It has been shown that in contrast to the static exchange framework there is some general global structure property of the equilibrium set on the one-parametrized level (Lehmann-Waffenschmidt 1983, 1985; Mas-Colell 1985; see also Chapter 10 in the present study). Actually, we can show more, namely that this is even the *only global* structure property which *generally* holds (Lehmann-Waffenschmidt 1988; see

Chapter 13 here). Taken together these findings achieve a complete characterization of the graph of the equilibrium correspondence of the exchange framework on the static and on the one-parametrized level.

Chapter 4 presents three well–known models of a pure exchange economy and puts them into the forms needed for our later analysis. In fact, the first model, borrowed from Arrow and Hahn (Section 4.1) contains the model by Dierker (Section 4.2) as a special case. Nevertheless, Dierker's version is also explicitly considered here since it allows for a different and more flexible formal treatment. It will later turn out to be considerably well–suited for generalization to other basic set–ups.

In his analysis of one-parametrized economies A. Mas-Colell (1985, Section 5.8) employs a certain model of a large exchange economy as basic set–up (see ibid., Sections 5.2, 5.4).[1] Here we employ a slightly different model of a large exchange economy (Section 4.3) which follows the lines of Dierker (1974, Chapter 12). The differences basically lie in the fact that Mas-Colell uses preferences and initial endowments as primitives, whereas Dierker's version directly builds on demand functions.

4.1 Evolutions Based on the Model of an Exchange Economy by Arrow and Hahn

(i) Let us start with a brief review of the well–known model of a pure exchange economy by Arrow and Hahn (1971, Chapter 2, particularly Sections 2.7, 2.8). There are n markets as it will be the case throughout the whole study. Excess demand is defined on the boundaryless (n–1)–dimensional price simplex $\mathring{\Delta}^{n-1}$ and possibly also in points from the boundary $\partial \Delta^{n-1}$. Furthermore, the excess demand function

$$\zeta : \Delta^{n-1} \setminus L \to \mathbb{R}^n$$
$$p \mapsto \begin{pmatrix} \zeta_1(p) \\ \vdots \\ \zeta_n(p) \end{pmatrix}$$

is bounded from below and is continuous on its domain $\Delta^{n-1} \setminus L$ where the exception set $L \subseteq \partial \Delta^{n-1}$ is an arbitrary closed subset of the boundary. ζ satisfies the budget identity (Walras' law) $p \cdot \zeta(p) = 0$ and the following well–known boundary condition which reflects desirability of each commodity:

given any sequence (p^k) in $\Delta^{n-1} \setminus L$ which converges to some $\bar{p} \in L$, then

$$\lim_{p^k \to \bar{p}} \sum_{i=1}^{n} \zeta_i(p^k) = +\infty$$

---

[1] When proceeding to the regular analysis, however, Mas–Colell (1985, Section 8.8) confines himself to finitely many agents.

(cf Assumption 6 (C') in Arrow/Hahn (1971), p. 31.)

From all these assumptions on ζ follows immediately that the last condition really means desirability for each commodity, i.e. that

$$\zeta_i(p^k) \to +\infty \quad \text{when} \quad (p_i^k) \to 0, i = 1, \ldots, n$$

To gain further economic insight let us quote here Arrow and Hahn's comment on their choice of the domain $\Delta^{n-1} \backslash L$ (1971, p. 21, last paragraph): "... this means that the demand for a free good ... [may be] bounded [if $L \neq \partial \Delta^{n-1}$]. Every individual becomes satiated with respect to any particular good. Unfortunately this assumption comes close to being inconsistent with the reasoning underlying Walras' law, which requires that at any point the household is unsatiated with respect to at least one good." (Square brackets by the author.)

In the pure exchange framework the *set of equilibria* is defined as the set of price vectors clearing all markets simultaneously, i.e. as the zero set of the excess demand function, $\zeta^{-1}(0^n) = \bigcap_{i=1}^n \zeta_i^{-1}(0)$. Arrow and Hahn, however, generalize this notion of equilibrium in the following way (1971, Chapter 2, Definition 1): a price vector $p \in \Delta^{n-1}$ is an *equilibrium (price vector)* for the economy ζ if $\zeta(p) \leq 0^n$. Note that this particularly means that excess supply on some, or even on all markets, is not inconsistent with equilibrium. This means that free disposal is implicitly assumed for all commodities. However, from the assumptions on ζ follows immediately that excess supply in equilibrium can only occur for free goods. Formally this means that if $\zeta(p) \leq 0^n$ and $\zeta_i(p) < 0$, then $p_i = 0$ (Arrow/Hahn 1971, Chapter 2, Theorem 1). Consequently, if for a given economy ζ equilibria with free goods do not occur then the set of equilibria in the sense of Arrow and Hahn equals the zero set of ζ.

If one desires to exclude equilibria with free goods – be it for economic or for technical reasons – the following additional mild assumption on ζ will obviulsy help:

choose an arbitrarily small $\epsilon > 0$. Then for any $p \in \Delta^{n-1}$ the relation $p_i < \epsilon$ implies that $\zeta_i(p) > 0$.

Clearly this assumption does not mean a severe restriction from the economic viewpoint. Consistency with the previous assumptions is evident.

Now we come to the crucial notion of an evolution of economies in our present set–up.

Definition 4.1. *An evolution (of economies) based on the model of an exchange economy by Arrow and Hahn, or for short, an* **exchange–I–evolution,** *is formally given by a C^0–uniformly continuous one-parametrization, or say one–parameter family, perturbation, or homotopy, of economies*

$$(\zeta_s)_{s\in[0,1]} : (\Delta^{n-1}\backslash L) \times [0,1] \to \mathbb{R}^n_+$$

$$(p,s) \mapsto \zeta_s(p) = \begin{pmatrix} \zeta_{1_s}(p) \\ \vdots \\ \zeta_{n_s}(p) \end{pmatrix}$$

from the homotopy price prism $(\Delta^{n-1}\backslash L) \times [0,1]$ *into the commodity space* $\mathbb{R}^n_+$. $s \in [0,1]$ *is called the state, or variation parameter. For any* $s \in [0,1]$ ζ_s *is called the* **s–state economy** *of the evolution* $(\zeta_s)_{s\in[0,1]}$, *and* ζ_{i_s} *is called the* **s–state excess demand function** *of commodity i. In accordance with intuition* ζ_0 *is also called* **initial state**, *or* **initial state economy**, *and* ζ_1 *the terminal state, or terminal economy, of the evolution* $(\zeta_s)_{s\in[0,1]}$.

We will see below that for our purposes the C^0–uniform continuity assumption can be replaced by the following 'uniformization' of the desirability condition: given any sequence (p^k, s^k) in $(\Delta^{n-1}\backslash L) \times [0,1]$ which converges to a point $(\overline{p},\overline{s}) \in L \times [0,1]$, then

$$\sum_{i=1}^{n} \zeta_{i_{s^k}}(p^k) \to +\infty .$$

We introduce this uniformized desirability condition for two main reasons. First, it is weaker than the C^0–uniform continuity assumption on the one-parametrization, and second it will later turn out to be most useful for the formalization of an evolution of large exchange economies (see Section 4.3 below).

Analogously to the static case the uniformized desirability condition prevents the equilibria of an evolution from coming arbitrarily close to those points of the boundary where the exess demand state functions are not defined. Figure 4.1 (= Figure 2.3) provides a simple one-dimensional example where this assumption is violated.

Convention *When addressing an evolution of economies henceforth we will call the underlying static model the* **basic model**, *or the* **basic set–up**.

It is easy to get a geometric intuition of the presented analytical formalization of an evolution of economies. Actually, each initial component function ζ_{i_0}, i.e. each single market excess demand function of the initial state economy $\zeta_0 = (\zeta_{1_0}, \ldots, \zeta_{n_0})^T$ is continuously deformed, or say perturbed, when the variation parameter $s \in [0,1]$ moves from 0 to 1. Or, to put it more formally, the continuous perturbation $(\zeta_{i_s})_{s\in[0,1]}$ is *continuously* governed by the scalar parameter $s \in [0,1]$.

To make our analysis compatible with concepts frequently used in the literature we emphasize the fact that our set of exchange–I–evolutions can be identified with a certain subset of the set of continuous paths in the space of exchange economies $\mathcal{E}_{ex}$ endowed with the usual product topology of C^0– uniform convergence on compacta.

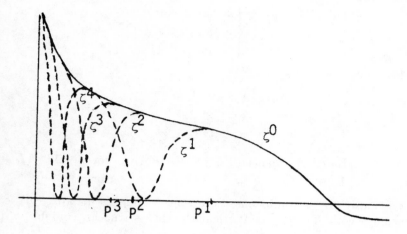

Fig. 4.1: Convergence Pathology (cf. Fig. 2.3)

Proposition 4.2. *An exchange–I–evolution* $(\zeta_s)_{s \in [0,1]} : (\Delta^{n-1} \backslash L) \times [0,1] \to$ $\mathbb{R}^n$ *canonically generates a continuous path of economies* $Z : [0,1] \to \mathcal{E}_{ex}$ *in the topological space of exchange economies* $\mathcal{E}_{ex}$ *by the canonical rule* $Z(s) := \zeta_s$. *Conversely, any continuous path* $Z : [0,1] \to \mathcal{E}_{ex}$ *of exchange economies generates an exchange-I–evolution* $(\zeta_s)_{s \in [0,1]}$ *if the one-parametrized family* $(\zeta_s)_{s \in [0,1]}$ *canonically generated by* $\zeta_s := Z(s)$ *satisfies the uniformized desirability assumption. Moreover, if we endow* $\mathcal{E}_{ex}$ *with the stronger topology of overall* C^0*–uniform convergence then any exchange–I–evolution generates a continuous path, and vice versa.*

We will postpone the proof of Proposition 4.2 until Chapter 5. There it will turn out to be a Corollary of the more general result of Proposition 5.2 below.

(ii) Now let us proceed to the second item on our agenda. This means that we have to provide a continuous self–mapping of the closed price simplex for any economy ζ such that the fixed point set equals the equilibrium set $\zeta^{-1}(0^n)$. We will call such a mapping an *equilibrium equivalent self–mapping* in the sequel.

To this end let us recall the well-known equilibrium equivalent self–mapping provided by Arrow and Hahn (1971, Section 2.8, pp. 31–32): first, define

$$K_\zeta(p) := \sum_{i=1}^{n} \zeta_i(p).$$ Now choose a continuous gluing function

$$\alpha : \mathbb{R} \to [0,1]$$

$$x \mapsto \begin{cases} 0 & \text{for } x \leq 0 \\ x & \text{for } 0 < x < 1 \\ 1 & \text{for } x \geq 1. \end{cases}$$

Let us define the mapping

$$N_\zeta : \Delta^{n-1} \to \mathbb{R}^n_+$$

by

$$N_{\zeta_i} : \Delta^{n-1} \to \mathbb{R}_+$$

$$p \mapsto \begin{cases} [1 - \alpha(K_\zeta(p)] \max(0, \zeta_i(p)) + \alpha(K_\zeta(p)) & \text{for } p \in \Delta^{n-1} \backslash L \\ 1 & \text{for } p \in L. \end{cases}$$

Obviously N_{ζ_i} is continuous. Note that N_{ζ_i} equals $\max(0, \zeta_i(p))$ for arguments p with $K_\zeta(p) \leq 0$. Particularly, this is true for equilibrium price vectors.

Now the zero problem of the existence of equilibrium price vectors is equivalently transformed into a fixed point problem by the following natural mapping

$$\varphi_\zeta : \Delta^{n-1} \to \Delta^{n-1}$$

$$p \mapsto \left(\frac{1}{\sum_{j=1}^n (p_j + N_{\zeta_j}(p))} \right) (p + N_\zeta(p))$$

(More detailed: $\varphi_{\zeta_i}(p) = \dfrac{p_i + N_{\zeta_i}(p)}{\sum_{j=1}^n (p_j + N_{\zeta_j}(p))}$.)

In fact, φ_ζ is an equilibrium equivalent self-mapping (see Arrow/Hahn, 1971, p. 32, for verifications).

(iii) Our last item fortunately is straightforward. From the construction of the equilibrium equivalent self–mapping φ_ζ and the definition of an exchange–I–evolution follows immediately that an exchange–I–evolution $(\zeta_s)_{s \in [0,1]}$ induces a *continuous one–parametrization of equilibrium equivalent self–mappings*

$$(\varphi_{\zeta_s})_{s \in [0,1]} : \Delta^{n-1} \times [0,1] \to \Delta^{n-1}$$

$$(p, s) \mapsto \varphi_{\zeta_s}(p).$$

4.2 Evolutions Based on Dierker's Version of the Model of an Exchange Economy

(i) Dierker's version of the model of an exchange economy (1974, Chapter 1) is a special case of the model by Arrow and Hahn. More precisely, the exception set L equals the whole boundary $\partial \Delta^{n-1}$. Nevertheless, as we have argued before from the economic viewpoint this case might be well considered as the most plausible one. Let us call an evolution based on Dierker's version of the model of an exchange economy an **exchange–II–evolution**.

Actually, what we here are primarily interested in is Dierker's approach to achieve a convenient equilibrium equivalent self–mapping. Actually, his construction will prove to be most useful for generalizations later in our analysis.

(ii) What we are going to present now is a slight modification of Dierker's original construction in (1974, proof of Theorem 8.3). Let an excess demand function (an economy)

$$\zeta : \mathring{\Delta}^{n-1} \to \mathbb{R}^n$$

be given. The equilibrium set is given by the zero set $\zeta^{-1}(0^n) = \bigcap_{i=1}^n \zeta_i^{-1}(0)$ of ζ. Consider now the derived mapping

$$f_\zeta : \mathring{\Delta}^{n-1} \to \mathbb{R}^n$$

$$p \mapsto \begin{pmatrix} p_1 + p_1 \zeta_1(p) \\ \vdots \\ P_n + p_n \zeta(p) \end{pmatrix}$$

Clearly, $\zeta^{-1}(0^n) = \mathrm{Fix}\,(f_\zeta)$. Note that Walras' law and boundedness from below imply that $p_i \zeta_i(p)$ is bounded from above for $i = 1, \ldots, n$. Thus f_ζ is bounded from below and from above as well. This will become important later. Due to Walras' law the image of f_ζ lies in the affine hyperplane H in $\mathbb{R}^n$ through Δ^{n-1}. For $h \in \{1, \ldots, n\}$ let us now define

$$V_h^\zeta := \left\{ p \in \mathring{\Delta}^{n-1} \,\middle|\, \zeta_h(p) > 0 \text{ and } p_h < \frac{1}{n} \right\}$$

$$V^\zeta := \bigcap_{h=1}^n V_h^\zeta,$$

and $\;K^\zeta := \mathring{\Delta}^{n-1} \backslash V^\zeta = (V^\zeta)^{co}.$

Evidently $\zeta^{-1}(0^n) \subseteq K^\zeta$. It is clear that K^ζ is closed in $\mathring{\Delta}^{n-1}$. Moreover, it is even compact in $\mathbb{R}^n$. To verify this we have to show that no sequence (p^k) in K^ζ converges to a point of the boundary $\partial \Delta^{n-1}$: for a sequence (p^k) in $\mathring{\Delta}^{n-1}$ with $p^k \to \partial \Delta^{n-1}$ the set $\mathcal{N}(p^k) := \{h \in \{1, \ldots, n\} | p_h^k \to 0\}$ is not empty. Assume now that there is no index $h \in \mathcal{N}(p^k)$ so that excess demand for commodity h becomes positive on a whole tail of the sequence (p^k). To put it more formally: for no $h \in \mathcal{N}(p^k)$ there exists a natural number $k(h)$ with $\zeta_h(p^k) > 0$ for *all* $k \geq k(h)$.

But then the desirability condition implies that there is at least one index $i \in \{1, \ldots, n\} \backslash \mathcal{N}(p^k)$ with

$$\zeta_i(p^k) \to +\infty.$$

Furthermore by Walras' law an index $j \in \{1, \ldots, n\}$ must exist with $\zeta_j(p^k) \to -\infty$. But this contradicts the assumption that ζ is bounded from below.

Consequently there is at least one index $h \in \mathcal{N}(p^k)$ and an index $k(h) \in \{1, \ldots, n\}$ so that $\zeta_h(p^k) > 0$ for all $k \geq k(h)$. But $h \in \mathcal{N}(p^k)$ implies that there is natural number $\ell(h)$ with $p_h^k \in V_h^\zeta$ for all $k \geq max(k(h), \ell(h))$. This means that $p^k \in V^\zeta$ for all $k \geq max(k(h), \ell(h))$. In other words, there is a tail (p^k) which is in V^ζ, and this proves the compactness of $K^\zeta = (V^\zeta)^{co}$.

In the next step we extend the mapping f_ζ to the boundary $\partial\Delta^{n-1}$ in a continuous way so that it remains unchanged on K^ζ and thus particularly also on $\zeta^{-1}(0^n)$. Since K^ζ is a compact subset of $\mathring{\Delta}^{n-1}$ the number

$$dist \ (K^\zeta, \partial\Delta^{n-1}) =: \gamma_\zeta$$

is well–defined and positive. Consider the inscribed closed simplex $\Delta_{\epsilon_\zeta}^{n-1}$ with $\epsilon_\zeta = (5/6)\gamma_\zeta$ which still contains K^ζ. Now choose a continuous 'hat function'

$$\lambda_\zeta : \Delta^{n-1} \to [0, 1]$$

with

$$\lambda_\zeta|_{\Delta_{\epsilon_\zeta}^{n-1}} \equiv 1 \ \text{ and } \ \lambda_\zeta|_{\partial\Delta^{n-1}} \equiv 0.$$

By means of λ_ζ we will glue f_ζ with the constant mapping which maps all points of the closed price simplex into its center $(1/n, \ldots, 1/n)$.
The resulting mapping

$$\widetilde{f}_\zeta : \Delta^{n-1} \to H \subset \mathbb{R}^n$$

$$p \mapsto \begin{cases} \lambda_\zeta(p)f_\zeta(p) + (1 - \lambda_\zeta(p)) \cdot (1/n, \ldots, 1/n) & \text{for } p \in \mathring{\Delta}^{n-1} \\ (1/n, \ldots, 1/n) & \text{for } p \in \partial\Delta^{n-1} \end{cases}$$

is continuous and equals f_ζ on the critical set K^ζ, i.e.

$$\widetilde{f}_\zeta|_{K^\zeta} = f_\zeta|_{K^\zeta}.$$

Now choose an arbitrary retraction of H on Δ^{n-1}, i.e. a continuous mapping $r : H \to \Delta^{n-1}$ with the property $r|_{\Delta^{n-1}} = id_{\Delta^{n-1}}$ and $r(H \backslash \Delta^{n-1}) = \partial\Delta^{n-1}$. We assert that the composition

$$r \circ \widetilde{f}_\zeta : \Delta^{n-1} \to \Delta^{n-1}$$

achieves the desired equilibrium equivalent selfmapping. To show this it only remains to verify that $Fix(r \circ \widetilde{f}_\zeta) = \zeta^{-1}(0^n)$.

Clearly $Fix \ (r \circ \widetilde{f}_\zeta) = Fix \ (\widetilde{f}_\zeta) \supset Fix(f_\zeta) = \zeta^{-1}(0^n)$. Thus, if we can show that $Fix \ (r \circ \widetilde{f}_\zeta) \subset K^\zeta$ we are done since $K^\zeta \subset \Delta_{\epsilon_\zeta}^{n-1}$ and

$$r \circ \widetilde{f}_\zeta|_{\Delta_{\epsilon_\zeta}^{n-1}} = \widetilde{f}_\zeta|_{\Delta_{\epsilon_\zeta}^{n-1}} = f_\zeta|_{\Delta_{\epsilon_\zeta}^{n-1}}.$$

To establish the inclusion $Fix(r \circ \widetilde{f}_\zeta) \subset K^\zeta$ choose an arbitrary $p \in Fix(\widetilde{f}_\zeta) = Fix(r \circ \widetilde{f}_\zeta)$. Consequently, $p \notin \partial \Delta^{n-1}$. Assume now that $p \notin K^\zeta$, i.e. there is an $h \in \{1, \ldots, n\}$ so that $p \in V_h^\zeta$. Then from the definition follows:

$$
\begin{aligned}
(\widetilde{f}_\zeta)_h(p) &= \lambda_\zeta(p)(\widetilde{f}_\zeta)_h(p) + (1 - \lambda_\zeta(p))1/n \\
&= \lambda_\zeta(p)(p_h + p_h \zeta_h(p)) + (1 - \lambda_\zeta(p))1/n \\
&> \lambda_\zeta(p)p_h + (1 - \lambda_\zeta(p))p_h \\
&= p_h.
\end{aligned}
$$

But this means $p \notin \mathrm{Fix}\, \widetilde{f}_\zeta$ in contradiction to our assumption. Hence, $p \in \mathring{\Delta}^{n-1} \setminus V^\zeta = K^\zeta$.

(iii) We are still left to verify that any exchange–II–evolution $Z := (\zeta_s)_{s \in [0,1]}$ induces a continuous one-parametrization of equilibrium equivalent self–mappings. From the C^0–uniform continuity of Z and the compactness of the prism $\Delta^{n-1} \times [0,1]$ follows immediately that one can choose a positive real number γ_Z with $0 < \gamma_Z < \mathrm{dist}(K^{\zeta_s}, \partial \Delta^{n-1})$ for all $s \in [0,1]$. Correspondingly to the construction above we choose a continuous hat function

$$\lambda_Z : \Delta^{n-1} \to [0,1]$$
$$\text{with} \quad \lambda_Z|_{\Delta_{\gamma_Z}^{n-1}} \equiv 1,$$
$$\text{and} \quad \lambda_Z|_{\partial \Delta^{n-1}} \equiv 0.$$

With the function λ_Z as gluing function the given exchange–II–evolution Z induces a continuous one-parametrization of equilibrium equivalent self–mappings as desired

$$(r \circ \overline{f}_{\zeta_s})_{s \in [0,1]} : \Delta^{n-1} \times [0,1] \to \Delta^{n-1}$$
$$(p, s) \mapsto (r \circ \overline{f}_{\zeta_s})(p)$$

where

$$\overline{f}_{\zeta_s} : \Delta^{n-1} \times [0,1] \to H$$
$$(p, s) \mapsto \begin{cases} \lambda_Z(p) \cdot f_{\zeta_s}(p) + (1 - \\ \lambda_Z(p))(1/n, \ldots, 1/n) & \text{for } (p, s) \in \mathring{\Delta}^{n-1} \times [0,1] \\ (1/n, \ldots, 1/n) & \text{for } (p, s) \in \partial \Delta^{n-1} \times [0,1]. \end{cases}$$

4.3 Evolutions Based on a Model of a Large Exchange Economy

(i) The last model from the traditional exchange framework which we will employ as a basic model in our study is a model of a *large exchange economy*. Actually, a model of a large exchange economy has also been used by Mas-Colell (1985, Chapter 5, especially Sections 5.2, 5.4.) as basic set–up for one-parametrized economies (Section 5.8, pp. 235-241). While Mas-Colell, however, uses individual preferences and endowments as primitives, here we basically will follow the lines of Dierker (1974, Chapter 12) who directly builds on demand functions (see also Hildenbrand 1974). Below we will discuss these differences in greater detail.

The space of characteristics of the agents is given by $D^0 \times (\mathbb{R}^n_+ \backslash \{0^n\})$. D^0 denotes the set of continuous individual demand functions

$$f : \mathring{\Delta}^{n-1} \times \mathbb{R}_{++} \to \mathbb{R}^n_+$$

which satisfy Walras' law: $p \cdot f(p, w) = w$ for all $(p, w) \in \mathring{\Delta}^{n-1} \times \mathbb{R}_{++}$, and a slightly strengthened desirability condition:

for any sequence (p^k, w^k) in $\mathring{\Delta}^{n-1} \times \mathbb{R}_{++}$ which converges to a pair $(p, w) \in \partial \Delta^{n-1} \times \mathbb{R}_{++}$ the norm $\| f(p^k, w^k) \|$ of the demand vectors grows beyond all finite bounds.

In order to ensure later that evolutions satisfy a *uniformized* desirability condition we still impose an additional boundary assumption on any individual demand function. For any positive real number β there is an (arbitrarily small) $\epsilon_\beta > 0$ such that for any $f \in D^0$

$$f_i(p, w) > \beta \quad \text{for any} \quad (p, w) \in \mathring{\Delta}^{n-1} \times \mathbb{R}_{++} \quad \text{with} \quad p_i < \epsilon_\beta.$$

$\mathbb{R}^n_+ \backslash \{0^n\}$ is the admissible space of initial endowments, and $\mathbb{R}_{++}$ is the space of individual wealth. Given an endowment bundle $\omega \in \mathbb{R}^n_+ \backslash \{0^n\}$ and a price vector $p \in \mathring{\Delta}^{n-1}$ individual wealth is given by the evaluation $\omega p \in \mathbb{R}_{++}$. $D^0 \times (\mathbb{R}^n_+ \backslash \{0^n\})$ is topologized by the product topology of uniform convergence on compacta (cf Dierker, 1974, p. 9, pp. 119–121).

The *space of large exchange economies* $\mathcal{M}_T$ is given by the set of all probability measures μ on $(T, B(T))$ where T is any non–empty compact subset of the characteristics space $D^0 \times (\mathbb{R}^n_+ \backslash \{0^n\})$, and $B(T)$ is the associated Borel algebra. Thus, a *large exchange economy* $\mu \in \mathcal{M}_T$ is completely characterized by the *distribution* of demand and initial endowment characteristics. Given an economy $\mu \in \mathcal{M}_T$ the *mean excess demand function* is the function

$$\zeta_\mu : \mathring{\Delta}^{n-1} \to \mathbb{R}^n$$
$$p \mapsto \zeta_\mu(p) = \int f(-, -, p)d\mu - \int \omega(-, -, p)d\mu$$

where

$$f : (D^0 \times \mathbb{R}^n_+ \backslash \{0^n\}) \times \mathring{\Delta}^{n-1} \to \mathbb{R}^n_+$$
$$(f, \omega; p) \mapsto f(p, \omega p)$$

is the natural evaluation mapping, and

$$\omega : (D^0 \times (\mathbb{R}_+^n \setminus \{0^n\})) \times \mathring{\Delta}^{n-1} \to \mathbb{R}_+^n \setminus \{0^n\}$$
$$(f, \omega; p) \mapsto \omega$$

is the projection on the second argument. Hence, the integrals $\int f(-, -, p)d\mu$ and $\int \omega(-, -, p)d\mu$ are in the usual sense vector–valued.

Note that the term $(-\int \omega(-, -, p)d\mu)$ is uniformly bounded from below on $\mathring{\Delta}^{n-1}$ since the economy subspace T is compact. Naturally, the set of equilibria of an economy μ is the set of zeroes of ζ_μ.

A simple example of a large economy shall be mentioned here: if T is finite, or if μ has a finite support with k elements, then the following mean excess demand function obtains: $\zeta_\mu(p) := \sum_{i=1}^k \alpha_i f^i(p, p\omega^i) - \sum_{i=1}^k \alpha_i \omega^i$ with $(\alpha_1, \ldots, \alpha_k) \in \mathring{\Delta}^{k-1}$. If μ, moreover, is equally distributed, then the so–called 'per–capita excess demand function' obtains: $\zeta_\mu(p) := 1/k \sum_{i=1}^k f^i(p, p\omega^i) - 1/k \sum_{i=1}^k \omega^i$.

Now let us topologize the space $\mathcal{M}_T$ of economies in a way so that similar economies also have similar mean excess demand. Actually, this is achieved by the topology of weak convergence of probability measures. Moreover, $\mathcal{M}_T$ becomes a compact, separable, and complete metric space (cf Dierker, 1974, p. 122). Consequently, the mapping

$$\zeta : \mathcal{M}_T \times \mathring{\Delta}^{n-1} \to \mathbb{R}^n$$
$$(\mu, p) \mapsto \zeta_\mu(p)$$

is continuous. Particularly, ζ_μ is continuous for any μ and has the usual properties of an excess demand function (see ibid.).

Trying to formalize the notion of an evolution of economies in this basic set–up shows that the model of a large exchange economy is considerably more abstract than the ordinary exchange model. In fact, we cannot any longer use the geometrically intuitive notion of perturbed behavior functions. Instead, we have to switch to the concept of a continuous path in the topological space of large economies (see Mathematical Preliminaries). Accordingly, let us formalize an **evolution of large exchange economies,** or an **exchange–III–evolution** for short, by a continuous path

$$(\zeta_s)_{s \in [0,1]} : [0, 1] \to \mathcal{M}_T$$
$$s \mapsto \mu_s.$$

(Lehmann-Waffenschmidt (1985, pp. 55–56) provides alternative characterizations of continuous paths of large exchange economies which help to make this notion economically more intuitive.)

Finally, let us come back to the aforementioned differences between the model of a large exchange economy adopted here and that employed by Mas-Colell (1985, Section 5.2, 5.4). Mas-Colell uses as space of agents' characteristics the economically more immediate space $\mathcal{M}_{b,sc}^0 \times \mathbb{R}_{++}^n$, where $\mathbb{R}_{++}^n$

represents the admissible space of agents' initial endowments, and $\mathcal{M}^0_{b,sc}$ is the space of individual preference relations on the commodity space $\mathbb{R}^n_{++}$ which are *continuous*, monotone, strictly convex, and which satisfy a suitable boundary condition (1985, 5.2, and Definition 2.3.16).[2] In this context an economy is specified as a map $E : I \to \mathcal{M}^0_{b,sc} \times \mathbb{R}^n_{++}$ where the set of agents' names I is finite or equal to the unit interval $[0,1]$. Thus, an *economy* is completely described by the number of agents and by the distribution ν of the agents' characteristics. Note that $(\nu(B) = \lambda(E^{-1}(B)$ for $I = [0,1]$ where λ is the Lebesgue measure on $\mathbb{R}$ and B is an element of the Borel algebra of $\mathcal{M}^0_{b,sc} \times \mathbb{R}^n_{++}$ (see Mas-Colell, 1985, p. 184).

Convention *Let us for the sake of better distinction subsequently call an economy in Mas-Colell's set–up an **explicit large exchange economy**. Accordingly, if the set of agents' names is finite we will speak of an **explicit finite exchange economy**.*

The space of explicit exchange economies M is endowed with the topology which is derived from the following natural concept of nearness: a sequence of economies $E_n : I_n \to \mathcal{M}^0_{b,sc} \times \mathbb{R}^n_{++}$ is converging to an economy $E : I \to \mathcal{M}^0_{b,sc} \times \mathbb{R}^n_{++}$ when (i) $1/\sharp I_n \to 1/\sharp I$, (ii) supp $\gamma_n \to$ supp γ in the Hausdorff distance, and (iii) $\gamma_n \to \gamma$ weakly (see Mas-Colell, 1985, Def. 5.4.1).

(ii) The provision of an equilibrium equivalent self–mapping for our basic model of a large exchange economy is straightforward. Actually, all necessary preparations have already been done in the preceding Subsection. The mean excess demand function $\zeta_\mu(-)$ of a large economy μ has the same properties as an excess demand function in Dierker's version, and can consequently be treated in the same way (see Dierker, 1974, p. 122).

(iii) Likewise it is entirely straightforward to check that an evolution of large exchange economies induces a *continuous* one-parametrization of equilibrium equivalent self–mappings. The composition

$$\overset{\circ}{\Delta}{}^{n-1} \times [0,1] \xrightarrow{id_{\overset{\circ}{\Delta}{}^{n-1}} \times (\zeta_s)_{s \in [0,1]}} \overset{\circ}{\Delta}{}^{n-1} \times \mathcal{M}_T \xrightarrow{tr} \mathcal{M}_T \times \overset{\circ}{\Delta}{}^{n-1} \xrightarrow{\zeta} \mathbb{R}^n$$
$$(p,s) \qquad \mapsto \qquad (p,\mu_s)\} \qquad \mapsto (\mu_s,p) \qquad \mapsto \zeta_{\mu_s}(p)$$

clearly is continuous and satifies the uniformized desirability condition as exchange-II–evolutions do it. The latter is due to the additional boundary assumption on the individual demand functions of D^0.

---

[2] Mas-Colell (1985) primarily considers preference spaces with differentiability properties (p. 168). However, to make it comparable to our analysis here we report the general continuous case.

5

Evolutions in an Exchange Equilibrium Framework Without Walras' Law and Homogeneity

It is appealing for a theoretical economist for several reasons to relax the traditional assumptions of Walras' law, i.e. of the budget identity, and of homogeneity of degree zero of the excess demand function. To begin with the purely economic aspects, Walras' law and the homogeneity property of the excess demand function clearly do not make sense anymore when one studies *a subsystem* of all conceivable markets, thus placing oneself in a *partial equilibrium framework* (cf. Schulz, 1985, introductory remarks).

Relaxing the budget identity will furthermore turn out to be essential for our analysis of evolving economies in Chapter 19. Apparently, the budget identity does not make any sense for our "frequency model" (see Section 19.2.2) which models demand and supply over time by time intervals between two demanded (supplied) commodity units. Relaxing homogeneity furthermore allows one to take into account inflationary effects and effects from currency reforms on excess demand hurting the assumption of freedom of money illusion.

However, there is still a second type of motive for studying such a general framework. Traditionally, existence proofs of general equilibrium essentially hinge on Walras' law and homogeneity. This raises the natural question how deeply the consistency of the model under consideration depends on these assumptions.

The two model versions of an exchange economy without Walras' law and homogeneity which we are going to present in this Chapter are in the spirit of Schulz's model (1985). Nevertheless, our second model (Section 5.2) generalizes Schulz's model in that it also allows for bounded demand for free goods. Actually, in the present context satiation with free goods no longer conflicts with the other assumptions. The main difference between Schulz's and our framework lies in the fact that we use a fixed–point approach for formal treatment whereas Schulz employs a degree theoretical approach. While the degree theoretical approach certainly has its merits the fixed–point approach will be crucial for our later results.

The question may arise whether the presented models do really general-ize the traditional Walrasian exchange model. Actually, it is not hard to see that the presented models can easily be expanded further by adding one more appropriate excess demand function so that a Walrasian economy of the Ar-row/Hahn type obtains (cf. Section 4.1). But this is precisely what we are not allowed to do. We have to cope with a given system of n markets, no more and no less. Playing the role of an "advocatus diaboli", however, we even ask the more serious question whether possibly there is an $n + 1$. market already contained in the given system of n markets which may balance the model in the Walras sense. Actually, we will find one, namely the "market" where all cross entries of units of account are registered which are associated with any planned excess demand or excess supply on the n commodity markets. But we will be able to show that the obtained system of $n + 1$ excess demand functions generally does not satisfy the boundary conditions of the Walrasian framework.

5.1 Evolutions Based on a Model of an Exchange Economy Without Walras' Law and Homogeneity

(i) A *basic economy without Walras' law and homogeneity of type I* is given by a continuous excess demand function

$$\zeta : \mathbb{R}^n_+ \longrightarrow \mathbb{R}^n$$

$$p \mapsto \begin{pmatrix} \zeta_1\,(p) \\ \vdots \\ \zeta_n\,(p) \end{pmatrix}$$

with the following properties:

(1) Let (p^m) be an arbitrary unbounded sequence in $\mathbb{R}^n_+$. Define $J(p^m) :=$ $\{i \in \{1, \ldots, n\} | lim_m\, p^m_i = +\infty$ *and* there is a $k_i \in \mathbb{N}$ such that for all $k \geq k_i$ one has $p^k_i/(\sum^n_{j=1} p^k_j) \geq \frac{1,1}{10n}\}$. (This means that an index i is in $J(p^m)$ when the i–th component sequence p^m_i converges to infinity "so fast" that a whole tail of the i–th component sequence $(p^m_i / \sum^n_{j=1} p_j)$ of the intersection points of the rays λp^m with $\Delta^{n-1}, \lambda > 0$, lies in the interval $[\frac{1,1}{10n}, \infty[$. Clearly, $J(p^m)$ is always non–empty.) Then the following holds: there is an index $\widetilde{i} \in J(p^m)$ and a $k_0 \in \mathbb{N}$ such that $\zeta_{\widetilde{i}}(p^k) < 0$ for all $k \geq k_0$.
(2) Choose an arbitrary $y \in \partial \mathbb{R}^n_+$. Then at least for one index $\widetilde{j} \in \{1, \ldots, n\}$ with $y_{\widetilde{j}} = 0$ the following is true: $\zeta_{\widetilde{j}}(y) > 0$.
(3) There is a positive real number b such that for all $i = 1, \ldots, n$ and any $p \in \mathbb{R}^n_+$ one has $\zeta_i(p) > -b$.

Let us comment on this definition. The choice of the positive orthant $\mathbb{R}^n_+$ as price space is natural since we have no homogeneity property. Possibly, the reader may feel some uneasiness about the zero–vector beeing not excluded from the price domain. Actually, the zero–vector has only been included here in order to lower notational expense and to simplify the formal analysis in the Subsections (ii) and (iii) below. We will see that it also can be excluded in the more general setting of the following Section 5.2. Property (3) simply expresses the fact that supply cannot grow beyond all finite bounds. Properties (1) and (2) characterize the boundary behavior of ζ. While Property (2) reflects an intuitive weak desirability property of each commodity, Property (1) deserves more attention. Though it may give the impression of being somewhat artificial, it, nevertheless, formalizes an economically reasonable idea: whenever the prices of some commodities become arbitrarily large, then at least one of these commodities will eventually be in excess supply. To paraphrase it in another way the second one of the two conditions defining the index subset $J(p^m)$ just rules out an index j when the j–th component sequence (p_j^m) of (p^m) converges "so slowly" to infinity that the corresponding j–th component sequence of the intersection points of the vectors p^m with the $(n-1)$–unit–simplex Δ^{n-1} does not eventually exceed $\frac{1,1}{10n}$.

The shaded area in the following illustrating Figure 5.1 shows the subspace of $\mathbb{R}_+^2$ which contains all sequences (p^m) with $J(p^m) = \{1,2\}$. Note that by

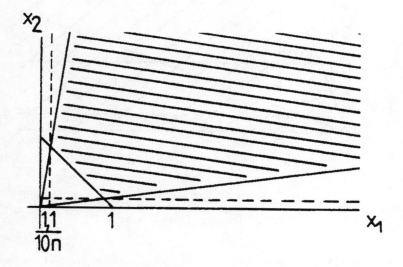

Fig. 5.1: Boundary Condition Areas I

Property (1) the market excess demand function *of at least one commodity i*, where i is an element of the distinguished index subset $J(p^m)$, is negative for the terms of a whole tail of the price sequence (p^m). Admittedly, the bound $\frac{1,1}{10n}$ seems to be somewhat arbitrary. However, it can be replaced, for example, by any number $\frac{1+\epsilon}{\alpha \cdot n}$ with an arbitrarily small positive real number ϵ and an arbitrarily large positive real number α. This follows immediately from the construction of the mapping $\widetilde{g}_{s_h}$ in Subsection (ii) below.

Apparently, the class of basic economies in this set–up is considerably large. To gain some intuition let us look at the following subclass: choose arbitrarily large positive real numbers α and β and an arbitrarily small positive real number $\gamma < \frac{1,1}{10n}$. A function $\zeta : \mathbb{R}^n_+ \longrightarrow \mathbb{R}^n$ is a basic economy when it meets the following conditions:

(a) if $p \in \mathbb{R}^n_+$ with $p_i > \alpha$ and $p_i/(\sum_{j=1}^n p_j) \leq \frac{1,1}{10n}$, then $-\beta \leq \zeta_i(p) < 0$.
(b) if $q \in \mathbb{R}^n_+$ with $q_i < \gamma$, then $\zeta_i(q) > 0$.

Figure 5.2 gives the geometric intuition. In the introductory remarks to this

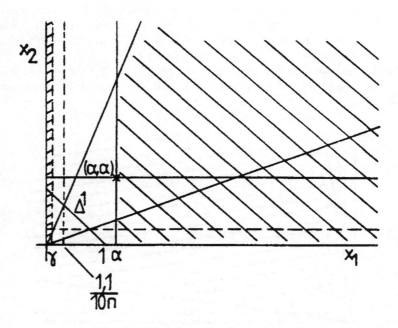

Fig. 5.2: Boundary Condition Areas II

Chapter we have given a general economic justification to consider a theoretical framework of pure exchange without Walras' law and homogeneity. Now we have to ensure that a basic economy in the present context is not essentially the same as a basic economy from the traditional Walrasian exchange framework.

To show this we will proceed in the following way. We start with playing the role of an "advocatus diaboli" trying to make a reasonable proposal to show the opposite. Afterwards we will see that his proposal does not work:

Let a basic economy $\zeta = \begin{pmatrix} \zeta_1 \\ \vdots \\ \zeta_n \end{pmatrix} : \mathbb{R}_+^n \longrightarrow \mathbb{R}^n$ with n commodity markets

from our present set–up be given. First we have to transform ζ into a mapping with the unit simplex as price domain in order to make it comparable with the traditional Walrasian set–up. We will achieve this by means of the two canonical homeomorphisms

$$\varphi : T^n \backslash \Delta^{n-1} \xrightarrow{\approx} \mathbb{R}_+^n,$$

$$p \mapsto \frac{1}{1 - \sum_{j=1}^n pj}\, p,$$

i.e. the central projection homeomorphism from the origin (see p.), and

$$\pi : \Delta^{(n+1)-1} \backslash \Delta^{n-1} \xrightarrow{\approx} T^n \backslash \Delta^{n-1},$$

$$(p_1, \ldots, p_n, p_{n+1}) \mapsto (p_1, \ldots, p_n),$$

i.e. the orthogonal projection (recall that $T^n = \{p \in \mathbb{R}_+^n | \sum_{i=1}^n p_i \leq 1\}$). In order to give the advocatus diaboli a chance let us for the moment consider of the composite mapping

$$\zeta \circ \varphi : T^n \backslash \Delta^{n-1} \longrightarrow \mathbb{R}^n.$$

Actually, this composition almost looks like a basic economy in the model by Arrow and Hahn with the domain exception set Δ^{n-1}. To fit $(\zeta \circ \varphi)$ still better into the set–up by Arrow and Hahn it appears natural to replace the domain $T^n \backslash \Delta^{n-1}$ by the homeomorphic unit simplex $\Delta^{(n+1)-1} \backslash \Delta^{n-1} \subset \mathbb{R}_+^{n+1}$. Now, we need an $n + 1$. market excess demand function which achieves Walras' law. Clearly, introducing a new commodity market from outside the given economy ζ would break the rules. However, there is a reasonable candidate for an $n + 1$. excess demand function which already is inherent in ζ. According to the sign convenience any planned excess demand $(\zeta_i \circ \varphi)(p) > 0$ (excess supply $(\zeta_i \circ \varphi)(p) < 0$) on some market i at price vector p naturally involves a nominal "cross entry" $- \; [(\zeta_i \circ \varphi)(p)] \cdot p_i < 0 \quad (-[(\zeta_i \circ \varphi)(p)] \cdot p_i > 0)$ of *excess supply (excess demand)* of *units of account*. Now let us take as $n + 1$. market the "market" where all supplies and demands of units of account caused by commodity excess demands and supplies are registrated. Thus, taking $p_{n+1} = 1$ as fixed price of one unit of account one can derive the following economically appealing mapping from the basic economy ζ :

$$(T^n \backslash \Delta^{n-1}) \times \{1\} \longrightarrow \mathbb{R}^{n+1}$$

$$\overline{\zeta} = \begin{pmatrix} \overline{\zeta}_1 \\ \vdots \\ \overline{\zeta}_n \\ \overline{\zeta}_{n+1} \end{pmatrix} := \begin{pmatrix} \zeta_1 \\ \vdots \\ \zeta_n \\ \overline{\zeta}_{n+1} \end{pmatrix} : \qquad (p,1) \mapsto \begin{pmatrix} (\zeta_1 \circ \varphi)(p) \\ \vdots \\ (\zeta_n \circ \varphi)(p) \\ -\sum_{i=1}^n [(\zeta_i \circ \varphi)(p)] \cdot p_i \end{pmatrix}$$

Obviously, $\overline{\zeta}$ fulfills Walras' law $\overline{\zeta}(p,1) \cdot (p,1) = 0$ for all $(p,1) \in (T^n \backslash \Delta^{n-1}) \times \{1\}$. Furthermore, the domain $(T^n \backslash \Delta^{n-1}) \times \{1\}$ is canonically homeomorphic to $\Delta^{(n+1)-1} \backslash \Delta^{n-1} \subset \mathbb{R}_+^{n+1}$ by means of the inverse of the orthogonal projection π. Consequently, $\overline{\zeta}$ strongly gives the impression of being a Walrasian exchange economy with $n+1$ markets and domain exception subset Δ^{n-1} in the set–up by Arrow and Hahn.

However, up to now we still have neglegted the boundary assumptions of the Arrow/ Hahn model. In fact, a brief calculation will show us that they are generally not satisfied by $\overline{\zeta}$. In the context of the price domain $(T^n \backslash \Delta^{n-1}) \times \{1\}$ used here the boundary assumption by Arrow and Hahn requires that for any sequence $(p^m, 1)$ in $(T^n \backslash \Delta^{n-1}) \times \{1\}$ which converges to some point $(p^\circ, 1) \in \Delta^{n-1} \times \{1\}$ the sum $\sum_{i=1}^{n+1} \overline{\zeta}_i(p^m)$ approaches infinity. By definition

$$\sum_{i=1}^{n+1} \overline{\zeta}(p^m) = \sum_{i=1}^{n} (\zeta_i \circ \varphi)(p^m) - \sum_{i=1}^{n} [(\zeta_i \circ \varphi)(p^m)] p_i^m.$$

However, from $0 \leq p_i^m \leq 1$ and from the boundedness from below of ζ follows that

$$+ \sum_{i=1}^{n} [(\zeta_i \circ \varphi)(p^m)] \cdot p_i^m$$

is bounded from below on $\Delta^{(n+1)-1} \backslash \Delta^{n-1}$. Consequently, the sum $\sum_{i=1}^{n+1} \overline{\zeta}_i(p^m)$ can reach infinity only if at least one of the n component sequences $((\zeta_i \circ \varphi)(p^m)), i = 1, \ldots, n$, does. But apparently, this need not be the case in general, and we are finished with the proof that a basic economy in the present context is essentially not the same as an economy from the traditional Walrasian framework.

After this digression we come to the precise notion of an evolution of economies in the present context.

Definition 5.1. *An **evolution of economies** based on the presented model of an exchange economy without Walras' law and homogeneity of type I, or an **exchange–III–evolution** for short, is a C°–uniformly continuous one-parametrization of basic economies*

$$(\zeta_s)_{s \in [0,1]} : \mathbb{R}_+^n \times [0,1] \longrightarrow \mathbb{R}^n.$$

As for the Walrasian case from Chapter 4 the C°–uniform assumption can be replaced by the following weaker *uniformized boundary condition:* for any sequence (p^m, s^m) in $\mathbb{R}_+^n \times [0,1]$ with an unbounded component sequence (p^m) and with (s^m) converging to some $s^\circ \in [0,1]$ there is at least one index $\widetilde{i} \in J(p^m)$ and one subsequence (p^{m_k}, s^{m_k}) such that for all natural k

$$\zeta_{\widetilde{i}_{s^{m_k}}}(p^{m_k}) < 0.$$

This assumption is a uniformization of Property (1) of a basic economy. Correspondingly to the uniformized desirability condition for Walrasian exchange evolutions it prevents the equilibria of exchange–III–evolutions from running to the boundary of the homotopy price space. Actually, it is not hard to find examples of continuous one-parametrizations which violate this assumption. However, from the economic viewpoint it is clearly not a restrictive assumption.

To make evolutions in the present context intuitive let us consider as an example the following continuous one-parametrizations

$$(\zeta_s)_{s\in[0,1]} : \mathbb{R}^n_+ \times [0,1] \longrightarrow \mathbb{R}^n$$

$$(p,s) \longmapsto \begin{pmatrix} (\hat{\zeta}_{1_s} \circ \varphi^{-1})(p) \\ \vdots \\ (\hat{\zeta}_{n_s} \circ \varphi^{-1})(p) \end{pmatrix}$$

where any component function $\zeta_{i_s} = \hat{\zeta}_{i_s} \circ \varphi^{-1} : T^n \longrightarrow \mathbb{R}$ meets the two following conditions:

(a) $\hat{\zeta}_{i_s}(p) > 0$ if $p_i > 0$,
(b) $\hat{\zeta}_{i_s}(p) < 0$ if $p \in \Delta^{n-1}$ and $p_j > \frac{1,1}{10n}$.

Again the reader is well–advised to geometrically envisage the functions $(\hat{\zeta}_{i_s} \circ \varphi^{-1})$. In Section 4.1, Proposition 4.2, we addressed the result on the equivalence of evolutions and paths of economies in the appropriately topologized space of Walrasian exchange economies.

Now we are going to prove the corresponding result in the present context without Walras' and law homogeneity:

Proposition 5.2. *Let $\mathcal{E}_{nwh}$ denote the space of economies in the present context endowed with the product topology of uniform convergence on compacta. An exchange–III–evolution*

$$(\zeta_s)_{s\in[0,1]}$$

canonically generates a continuous path $z : [0,1] \longrightarrow \mathcal{E}_{nwh}$ by the rule $z(s) := \zeta_s$. Conversely, if $z : [0,1] \longrightarrow \mathcal{E}_{nwh}$ is a continuous path such that the one-parametrization $(\zeta_s)_{s\in[0,1]}$ generated by the canonical rule $\zeta_s := z(s)$ satisfies the uniformized boundary assumption, then $(\zeta_s)_{s\in[0,1]}$ is continuous, i.e. is an exchange–III–evolution. Moreover, if we endow the space of economies $\mathcal{E}_{nwh}$ with the stronger topology of overall C°–uniform convergence, then an exchange–III–evolution generates a continuous path in $\mathcal{E}_{nwh}$, and vice versa.

Proof. We have first to show (1) that z is a continuous mapping, and (2) that $(\zeta_s)_{s\in[0,1]}$ is a continuous one-parametrization which satisfies the additional global boundary assumption.

(1) Let us show for an arbitrarily given exchange–III–evolution $(\zeta_s)_{s\in[0,1]}$ that $z : s \mapsto \zeta_s$ is a continuous mapping into $\mathcal{E}_{nwh}$. This means we have to show

that for any convergent sequence $s^m \longrightarrow s^o$ in $[0,1]$ the sequence ζ_{s^m} converges to ζ_{s^o} in $\mathcal{E}_{nwh}$ with respect to the product topology of uniform convergence on compact sets, i.e.,

$$\forall_{A \subset \mathbb{R}_+^n \text{ compact}} \forall_{\epsilon > 0} \exists_{l_\epsilon \in \mathbb{N}} \forall_{l \geq l_\epsilon} \max_{i \in \{1,\ldots,n\}} \max_{p \in A} |\zeta_{i_{s^l}}(p) - \zeta_{i_{s^o}}(p)| < \epsilon.$$

Let us *assume that z is not continuous*. This means:

$$\exists_{A \subset \mathbb{R}_+^n \text{ compact}} \exists_{\epsilon_A > 0} \exists_{\text{subsequence}} (s^{m_k}) \max_{i \in \{1,\ldots,n\}} \max_{p \in A} |\zeta_{i_{s_k^m}}(p) - \zeta_{i_{s^o}}(p)| \geq \epsilon.$$

Thus, there is at least one index $j \in \{1,\ldots,n\}$ and a subsequence $(s^{m_{\tilde{k}}})$ of (s^{m_k}) so that for each $m_{\tilde{k}}$ a point $p^{m_{\tilde{k}}} \in A$ can be choosen which realizes the double maximization i.e.

$$max_{i \in \{1,\ldots,n\}} max_{p \in A} |\zeta_{i_{s} m_{\tilde{k}}}(p) - \zeta_{i_{s^o}}(x)| = |\zeta_{j_{s_k^m}}(p^{m_{\tilde{k}}}) - \zeta_{j_{s^o}}(p^{m_{\tilde{k}}})| \geq \epsilon.$$

As $A \times [0,1]$ is compact there is a subsequence $(p^{m'}, s^{m'})$ of $(p^{m_{\tilde{k}}}, s^{m_{\tilde{k}}})$ which converges to a point $(p^o, s^o) \in A \times [0,1]$. Clearly $\epsilon_A \leq |\zeta_{j_{s^o}}(p^{m'}) - \zeta_{j_{s_{m'}}}(p^{m'})| \leq |\zeta_{j_{s^o}}(p^{m'}) - \zeta_{j_{s^o}}(p^o)| + |\zeta_{j_{s^o}}(p^o) - \zeta_{j_{s_{m'}}}(p^{m'})|$ for any m'. The first summand $|\zeta_{j_{s^o}}(p^{m'}) - \zeta_{j_{s^o}}(p^o)|$ of the right side *converges to zero* since $\zeta_{j_{s^o}}$ is continuous and $p^{m'} \longrightarrow p^o$. Hence, $|\zeta_{j_{s^o}}(p^o) - \zeta_{j_{s_{m'}}}(p^{m'})|$ *cannot converge to zero*. However, this contradicts our assumption that particularly the j–th market evolution $(\zeta_{j_s})_{s \in [0,1]} : \mathbb{R}_+^n \times [0,1] \longrightarrow \mathbb{R}$ is continuous at (p^o, s^o). Thus the continuity of the path z is proven.

(2) It remains to show that $(\zeta_s)_{s \in [0,1]} : \mathbb{R}_+^n \times [0,1] \longrightarrow \mathbb{R}^n$ is *continuous* if $z : s \mapsto \zeta_s$ is continuous. Choose a converging sequence $(p^m, s^m) \longrightarrow (p^o, s^o) \in \mathbb{R}_+^n \times [0,1]$. We have to show that $\zeta_{i_{s^m}}(p^m) \longrightarrow \zeta_{i_{s^o}}(p^o)$ for every $i \in \{1,\ldots,n\}$. Choose a compact subset $K \subset \mathbb{R}_+^n$ which contains all terms of the sequence (p^m). Consider the following inequality which is true for every i :

$$|\zeta_{i_{s^o}}(p^o) - \zeta_{i_{s^m}}(p^m)| \leq |\zeta_{i_{s^o}}(p^o) - \zeta_{i_{s^o}}(p^m)| + |\zeta_{i_{s^o}}(p^m) - \zeta_{i_{s^m}}(p^m)|.$$

Choose an arbitrary $i \in \{1,\ldots,n\}$ and an arbitrary $\epsilon > 0$. There is a whole tail $(p^{m'})$ of (p^m) so that $|\zeta_{i_{s^o}}(p^o) - \zeta_{i_{s^o}}(p^{m'})| < \frac{\epsilon}{2}$ since ζ_{s^o} is continuous. Furthermore, there is a tail $(p^{m''}, s^{m''})$ of (p^m, s^m) so that $|\zeta_{i_{s^o}}(p^{m''}) - \zeta_{i_{s_{m''}}}(p^{m''})| < \frac{\epsilon}{2}$ since z is continuous. This proves the continuity of the one-parametrization $(\zeta_s)_{s \in [0,1]}$.

The last statement of Proposition 5.2 follows directly as a simple special case from the above proof.   □

Evidently the proof of Proposition 5.2 directly carries over to Proposition 4.2 from Section 4.1.

(ii) Now we are going to provide an equilibrium equivalent self–mapping for our present basic model of an exchange economy without Walras' law and homogeneity of type I. This time Dierker's constructions in (1974, proof of Theorem 8.3) only serve as a rough guideline.

We start with an arbitrary basic economy ζ_s with fixed $s \in [0,1]$. We introduce the state index s already here in order to make the constructions in Subsection (iii) below more intuitive. Note that the equilibrium set of ζ_s

$$G_{\zeta_s} = \zeta_s^{-1}(0)$$

equals the set $\varphi[(\zeta_s \circ \varphi)^{-1}(0^n)]$.

This equation holds because (1) for $p \in \zeta_s^{-1}(0^n)$ one gets $(\zeta_s \circ \varphi)(\varphi^{-1}(p)) = 0^n$, and (2) the relation $y \in \varphi[(\zeta_s \circ \varphi)^{-1}(0^n)]$ can be transformed into $\varphi^{-1}(y) \in (\zeta_s \circ \varphi)^{-1}(0^n) \Leftrightarrow (\zeta_s \circ \varphi)(\varphi^{-1}(y)) = 0^n$. As φ is a homeomorphism, the equilibrium analaysis, i.e. the zero analysis, of ζ is equivalent to the zero analysis of $\zeta \circ \varphi$. Moreover, we can transform the *zero problem into a fixed–point problem* by means of the mapping

$$g_s = \zeta_s \circ \varphi + id : \overline{T}^n \backslash \Delta^{n-1} \longrightarrow \mathbb{R}^n$$
$$p \mapsto (\zeta_s \circ \varphi)(p) + p.$$

Certainly,

$$Fix\ (g_s) = (\zeta_s \circ \varphi)^{-1}(0^n).$$

Thus

$$G_{\zeta_s} = \zeta_s^{-1}(0^n) = \varphi[(\zeta_s \circ \varphi)^{-1}(0^n)] = \varphi[Fix\ (g_s)].$$

Consequently, the zero problem in $\mathbb{R}_+^n$ has been drawn back to $\overline{T}^n \backslash \Delta^{n-1}$ and has been equivalently transformed into a fixed–point problem.

We need two preparatory technical results on the boundary behavior of ζ_s.

(1) For any $z \in \overline{T}^n$ define the index subset $H(z) := \{i | z_i > \frac{1}{10n}\} \subset \{1, \ldots, n\}$. Choose an arbitrary $p \in \Delta^{n-1}$. (Of course, $H(p) \neq \emptyset$.) Then there exists an $\epsilon_p > 0^n$ so that the following holds:

$$\forall_{y \in \overset{\circ}{B}^n_{\epsilon_p}(p) \cap \overset{\circ}{T}^n}[H(y) \supseteq H(p)\ \text{and}\ \exists_{i_y \in H(y)}(\zeta_s \circ \varphi)_{i_y}(y) < 0].$$

Proof.   Clearly there is an $\epsilon'_p > o$ so that the following holds:

$$\forall_{y \in \overset{\circ}{B}^n_{\epsilon'_p}(p) \cap \overset{\circ}{T}^n} H(y) \supseteq H(p).$$

We still have to show that there exists an $\epsilon_p > 0$ $(\epsilon_p \leq \epsilon'_p)$ so that the following is true:

$$\forall_{y \in \overset{\circ}{B}^n_{\epsilon_p}(p) \cap \overset{\circ}{T}^n} \exists_{i_y \in H(y)} (\zeta_s \circ \varphi)_{i_y}(y) = ((\zeta_s)_{i_y}(\varphi(y)) < 0.$$

Suppose there is no $\epsilon_p > 0$ with this property. Then in every relatively open neighbourhood $\overset{\circ}{B}^n_{1/m}(p) \cap \overset{\circ}{T}^n$ of $p, m \in \mathbb{N}$ and $m \geq 1/\epsilon'_p$, there is a y^m with the following property:

$(*)$ $\forall_{i \in H(y^m)}(\zeta_s \circ \varphi)_i(y^m) \geq 0.$

Obviously (y^m) is a sequence in $\overset{\circ}{T}{}^n$ converging to p. Furthermore, there exists a subsequence (y^{m_k}) of (y^m) so that the following holds:

$$\forall_{k',k'' \in \mathbb{N}, k' \neq k''} H(y^{m_{k'}}) = H(y^{m_{k''}}) \supseteq J(\varphi(y^{m_k})).$$

(Clearly, the image sequence $(\varphi(y^{m_k}))$ is unbounded.)

To prove the last assertion note that $H(y^m)$ is non–empty for the terms of a tail of (y^m) and that furthermore there are only finitely many different subsets of the set $\{1, \ldots, n\}$. Thus, there is a subsequence $(y^{m_{\tilde{k}}})$ of (y^m) so that the index subsets $H(y^{m_{\tilde{k}}})$ and $H(y^{m_{\tilde{l}}})$ are equal for any two different indices $\tilde{k}$ and $\tilde{l}$.

Consider the sequence $(z^{m_{\tilde{k}}})$ of the intersection points of the rays $\lambda \cdot y^{m_{\tilde{k}}}, \lambda > 0$, with Δ^{n-1} (i.e., $z^{m_{\tilde{k}}} = (y^{m_{\tilde{k}}})/(\sum_{j=1}^n y_j^{m_{\tilde{k}}})$). Since $(y^{m_{\tilde{k}}})_{\tilde{k}}$ converges to p, consequently also $(z^{m_{\tilde{k}}})$ converges to p, and $J(\varphi(y^{m_{\tilde{k}}}))$ is non–empty. This yields the following implication:

$$j \in J(\varphi(y^{m_{\tilde{k}}})) \Rightarrow p_j \geq \frac{1,1}{10n}.$$

Thus, there exists a tail $(y^{m_{\widetilde{k'}}})$ of $(y^{m_{\tilde{k}}})$ with $y_j^{m_{\widetilde{k'}}} > \frac{1}{10n}$ for any $k' \in \mathbb{N}$. This means that $\forall_{k' \in \mathbb{N}}\, j \in H(y^{m_{\widetilde{k'}}})$. Since $J(\varphi(y^{m_{\tilde{k}}}))$ is finite, there exists an intersection tail (y^{m_k}) of $(y^{m_{\tilde{k}}})$ so that $J(\varphi(y^{m_k})) \subseteq H(y^{m_k})$ for any $\in \mathbb{N}$, and the proof is finished.

Consequently, supposition $(*)$ from above implies

$$\forall_{i \in J(\varphi(y^{m_k}))}\forall_{k \in \mathbb{N}}\, \zeta_{i_s}(\varphi(y^{m_k})) \geq 0^n.$$

But Property (1) of the economy ζ_s guarantees the existence of a tail of $(\varphi(y^{m_k}))$ so that the market excess demand function for at least one commodity i with $i \in J(\varphi(y^{m_k}))$ is negative for the terms of this tail.

This contradiction proves assertion (1).

(2) Choose an arbitrary $y \in \partial \overline{T}^n \backslash \Delta^{n-1}$. Then there exists an $\epsilon_y \in \,]0, \frac{1}{15n}[$ and a $\delta > 0$ such that

$$\exists_{i_y} \text{ with } y_{i_y}=0 \forall_{z \in \overset{\circ}{B}{}^n_{\epsilon_y}(y) \cap (\overline{T}^n \backslash \Delta^{n-1})} (\zeta_s \circ \varphi)_{i_y}(z) > \delta.$$

This follows immediately from Property (2) and from the continuity of ζ_s.

From our efforts we have achieved a relatively open covering U_o of ∂T^n in T^n:

$$U_o := (\cup_{p \in \Delta^{n-1}} \overset{\circ}{B}{}^n_{\epsilon_p}(p) \cap \overline{T}^n) \cup (\cup_{y \in \partial \overline{T}^n \backslash \Delta^{n-1}} \overset{\circ}{B}{}^n_{\epsilon_y}(y) \cap \overline{T}^n).$$

Clearly ∂T^n is compact. Therefore, finitely many of these sets are sufficient to cover ∂T^n :

$$\partial T^n \subset U_1 := \cup_{v_i \in \{v_1, \ldots, v_k\}} \overset{\circ}{B}{}^{\,n}_{\epsilon_{v_i}} (v_i) \cap \overline{T}^n.$$

Of course,

$$\theta := \frac{dist \, [\partial \overline{T}^n, \overline{T}^n \backslash U_1]}{2} > 0.$$

Define $\gamma := min \, \{\theta, \frac{1}{20n}\}$. Let us now consider $T^{n,1-\gamma}_\gamma$ which is inscribed into T^n and whose faces are parallel to those of $\overline{T}^n$ with distance γ. Apparently, U_1 is a relatively open covering of $\overline{T}^n \backslash \overline{T}^{n,1-\gamma}_\gamma$ in $\overline{T}^n$. Furthermore, the above statements (1) and (2) clearly imply that $(\overline{T}^n \backslash \overline{T}^{n,1-\gamma}_\gamma) \cap (\zeta_s \circ \varphi)^{-1}(0)$ is empty. Note that the point $M = (\frac{1}{10n}, \ldots, \frac{1}{10n})$ lies in the interior of $\overline{T}^{n,1-\gamma}_\gamma$. It will be crucial for the following construction.

Choose a continuous separating function

$$\lambda : \overline{T}^n \longrightarrow [0,1] \;\; \text{with}$$
$$\lambda|_{\overline{T}^{n,1-\gamma}_\gamma} \equiv 1$$
$$\lambda|_{\partial \overline{T}^n} \equiv 0^n.$$

Let us glue $g_s = (\zeta_s \circ \varphi + id)$ together with the constant mapping

$$M : \overline{T}^n \longrightarrow M$$

by means of λ. Thus we obtain a *continuous mapping*

$$\tilde{g}_s : \overline{T}^n \longrightarrow \mathbb{R}^n$$

defined by

$$\tilde{g}_{s_h}(p) = \begin{cases} \lambda(p) \cdot [(\zeta_{s_h} \circ \varphi)(p) + p_h] + (1 - \lambda(p))\frac{1}{10n}, & \text{for } p \in \overset{\circ}{T}{}^{\,n}, \\ 1/10n, & \text{for } p \in \partial T^n \end{cases}$$

$h = 1, \ldots, n$. Obviously, $\tilde{g}_s|_{\overline{T}^{n,1-\gamma}_\gamma} = \zeta_s \circ \varphi + id$ and $\tilde{g}_s|_{\partial \overline{T}^n} = M$. $\tilde{g}_s$ has the following crucial property:

$$Fix \, (\tilde{g}_s) \cap (\overline{T}^n \backslash \overline{T}^{n,1-\gamma}_\gamma) = \emptyset.$$

Let us prove the last assertion. For $y \in \partial \overline{T}^n$ the image $\tilde{g}_s(y)$ is precisely the point M. Therefore the only possible candidates for fixed–points are the points of $\overset{\circ}{T}{}^{\,n} \backslash \overline{T}^{n,1-\gamma}_\gamma$. Choose an arbitrary $p \in \overset{\circ}{T}{}^{\,n} \backslash \overline{T}^{n,1-\gamma}_\gamma$. $U_1 \supset \overline{T}^n \backslash \overline{T}^{n,1-\gamma}_\gamma$ implies furthermore that there is a $v_i \in \{v_1, \ldots, v_k\}$ so that

$$p \in \overset{\circ}{B}{}^{\,n}_{\epsilon_{v_i}} (v_i) \cap \overset{\circ}{T}{}^{\,n}.$$

According to the statements (1) and (2) above, *at least one of the following two statements is true:*

($\bar{1}$) $\exists_{\bar{i} \in \{1,\ldots,n\}} (\zeta_s \circ \varphi)_{\bar{i}}(p) < 0$, and $\frac{1}{10n} < p_{\bar{i}}$.

($\bar{2}$) $\exists_{\bar{j} \in \{1,\ldots,n\}} (\zeta_s \circ \varphi)_{\bar{j}}(p) > 0$, and $\frac{1}{10n} > x_{\bar{j}}$.

Since $\widetilde{g}_{s_h}(p) = \lambda(p)[(\zeta_s \circ \varphi)_h(p) + p_h] + (1 - \lambda(p))\frac{1}{10n}$, the following implications immediately obtain:

$$(\bar{1}) \Rightarrow \widetilde{g}_{s_{\bar{i}}} < p_{\bar{i}}, \quad \text{and}$$
$$(\bar{2}) \Rightarrow \widetilde{g}_{s_{\bar{j}}}(p) > p_{\bar{j}}.$$

Consequently, p is not a fixed–point of $\widetilde{g}_s$, and the assertion is proved.

Thus, $Fix(\widetilde{g}_s) = Fix(g_s)$.

Finally we choose an arbitrary retraction $r : \mathbb{R}^n \longrightarrow \overline{T}^n$ of $\mathbb{R}^n$ onto $\overline{T}^n$ with

$$r|_{\overline{T}^n} = id_{\overline{T}^n},$$

and

$$r(\mathbb{R}^n \setminus \overset{\circ}{T}{}^n) = \partial \overline{T}^n.$$

Obviously the self–mapping

$$r \circ \widetilde{g}_s : \overline{T}^n \longrightarrow \overline{T}^n$$

has no fixed–points in $\overline{T}^n \setminus \overline{T}_\gamma^{n,1-\gamma}$. Consequently, $Fix(r \circ \widetilde{g}_s) \subset \overline{T}_\gamma^{n,1-\gamma}$. But

$$Fix(r \circ \widetilde{g}_s) = Fix(\widetilde{g}_s) = Fix(g_s),$$

since

$$r \circ \widetilde{g}_s|_{\overline{T}_\gamma^{n,1-\gamma}} = \widetilde{g}_s|_{\overline{T}_\gamma^{n,1-\gamma}} = g_s|_{\overline{T}_\gamma^{n,1-\gamma}}.$$

This means, $r \circ \widetilde{g}_s$ is an equilibrium equivalent self–mapping.

There is a last remark in order: the preceding constructions make it evident that the bound $\frac{1,1}{10n}$ in Property (1) of a basic economy may equally be replaced by $\frac{1+\epsilon}{\alpha n}$, where α is an arbitrarily large and ϵ an arbitrarily small positive real number. Then $M = (1/10n, \ldots, 1/10n)$ has to be replaced by $M' = (\frac{1}{\alpha n}, \ldots, \frac{1}{\alpha n})$, γ by $\gamma' = min\,(\vartheta, \frac{1}{2\alpha n})$, and ϵ_y in the statement of the technical result (2) by $\epsilon_y \in]0, \frac{1}{1,5\alpha n}[$.

(iii) It remains to verify that an exchange–III–evolution $(\zeta_s)_{s \in [0,1]}$ induces a continuous one-parametrization $(r \circ \widetilde{g}_s)_{s \in [0,1]}$ of the corresponding equilibrium equivalent self–mappings. To show this let us first extend the statements (1) and (2) from Subsection (ii) to the whole evolution $(\zeta_s)_{s \in [0,1]}$.

(1') Choose an arbitrary $(p, \bar{s}) \in \Delta^{n-1} \times [0,1] \subset \partial \overline{T}^n \times [0,1]$. Then there exists an $\epsilon_{(p,\bar{s})} > 0$ so that the following holds:

$$\forall_{(y,s) \in \overset{\circ}{B}{}^{n+1}_{\epsilon_{(p,\bar{s})}}(p,\bar{s}) \cap \overset{\circ}{T}{}^n \times [0,1]} [H(y) \supseteq H(p) \text{ and } \exists_{i_{(y,s)} \in H(y)} (\zeta_s \circ \varphi)_{i_{(y,s)}}(y) < 0].$$

Proof. As the additional component s does not matter for $H(y)$ and $H(p)$, it remains to show in complete analogy to the unparametrized case from Subsection (ii):

$$\exists_{\epsilon''_{(p,\bar{s})}>0} \forall_{(y,s)\in \overset{\circ}{B}{}^{\,n+1}_{\epsilon''_{(p,\bar{s})}}(p,\bar{s})\cap \overset{\circ}{T}{}^{\,n}} \exists_{i_{(y,s)}\in H(y)} (\zeta_s \circ \varphi)_{i_{(y,s)}}(y) < 0.$$

Assume again that there is no $\epsilon''_{(p,\bar{s})}$ with this property. Thus, one finds a (y^m, s^m) in every neighbourhood $\overset{\circ}{B}{}^{\,n+1}_{1/m}(p,\bar{s})\cap \overset{\circ}{T}{}^{\,n}$, $m \in \mathbb{N}$, with the property

$$\forall_{i\in H(y^m)}(\zeta_s^m \circ \varphi)_i(y^m) = \zeta_{i_s m}(\varphi(y^m)) \geq 0.$$

Consider the component sequence (y^m). According to the considerations from the proof of (1) in Subsection (ii) there is a subsequence (y^{m_k}) with the property that all index subsets $H(y^{m_k})$ are identical and, furthermore, contain $J(\varphi(y^{m_k}))$.

Consider now the sequence (y^{m_k}, s^{m_k}). The above assumption implies

$$\forall_{i\in J(\varphi(y^{m_k}))}\forall_{k\in \mathbb{N}} \zeta_{i_s m_k}(\varphi(y^{m_k})) \geq 0.$$

However, this contradicts the uniformized boundary property of the evolution $(\zeta_s)_{s\in[0,1]}$.

(2') Choose an arbitrary $(y,\tilde{s}) \in (\partial \overline{T}{}^{\,n}\backslash\Delta^{n-1}) \times [0,1]$. Then there exists an $\epsilon_{(y,\tilde{s})} > 0$ so that

$$\exists_{i_{(y,\tilde{s})}\in\{1,\ldots,n\}} \forall_{z\in \overset{\circ}{B}{}^{\,n+1}_{\epsilon_{(y,\tilde{s})}}(y,\tilde{s})\cap \overline{T}{}^{\,n}\times[0,1]} \zeta_{i_{(y,\tilde{s})}}(z) > 0.$$

Again, this is clear from Property (2) of the basic economy $\zeta_{\tilde{s}}$ and the continuity of $(\zeta_s)_{s\in[0,1]}$.

In complete analogy to the unparametrized case treated in Subsection (ii) one can find a finite relatively open covering U'_1 of the "prism jacket" $\partial \overline{T}{}^{\,n} \times [0,1]$ in the prism $\overline{T}{}^{\,n} \times [0,1]$,

$$U'_1 := \cup_{(w_i,t_i)\in\{(w_1,t_1),\ldots,(w_l,t_l)\}}(\overset{\circ}{B}{}^{\,n+1}_{\epsilon_{(w_i,t_i)}}(w_i,t_i)) \cap (\overline{T}{}^{\,n} \times [0,1]).$$

Like before we define $\gamma' := min\,\{\frac{dist\,[\partial\overline{T}{}^{\,n}\times[0,1],(\overline{T}{}^{\,n}\times[0,1])\backslash U'_1]}{2}, \frac{1}{20n}\}$. Thus $((\overline{T}{}^{\,n} \times [0,1])\backslash(\overline{T}{}^{\,n,1-\gamma'}_{\gamma'} \times [0,1])) \cap (\cup_{t\in[0,1]}(\zeta_t \circ \varphi)^{-1}(0)) = \emptyset$. This means that the equilibrium price vectors of the whole exchange–III–evolution $(\zeta_s)_{s\in[0,1]}$ are *uniformly kept away from the prism jacket* $\partial\overline{T}{}^{\,n} \times [0,1]$.

In complete analogy to the preceding Subsection let us choose a separating functional

$$\lambda_Z : \overline{T}{}^{\,n} \longrightarrow [0,1]$$

with

$$\lambda_Z|_{\overline{T}_{\gamma'}^{n,1-\gamma'}} \equiv 1$$
$$\lambda_Z|_{\partial \overline{T}^n} \equiv 0.$$

Hence, we can define the continuous one-parametrization

$$(\widetilde{g}_s)_{s\in[0,1]} : \overline{T}^n \times [0,1] \longrightarrow \mathbb{R}^n$$
$$(x,s) \longmapsto \widetilde{g}_s(x)$$

by

$$\widetilde{g}_{s_h}(p) := \lambda_Z(p)[(\zeta_{s_h} \circ \varphi)(p) + p_h] + (1 - \lambda_Z(p))\frac{1}{10n}$$
$$= \lambda_Z(p)(g_{s_h}(p)) + (1 - \lambda_Z(p))\frac{1}{10n}, \qquad h = 1,\ldots,n.$$

In complete analogy to Subsection (ii) one sees that the one-parametrization $(\widetilde{g}_s)_{s\in[0,1]}$ leads to the continuous one-parametrization $(r \circ \widetilde{g}_s)_{s\in[0,1]}$: $\overline{T}^n \times [0,1] \longrightarrow \overline{T}^n$ of equilibrium equivalent self–mappings of $\overline{T}^n$.
From the considerations in (ii) follows directly that

$$\cup_{s\in[0,1]}Fix\ (g_s) = \cup_{s\in[0,1]}Fix\ (\widetilde{g}_s) = \cup_{s\in[0,1]}Fix\ (r \circ \widetilde{g}_s) \subset \overline{T}_{\gamma'}^{n,1-\gamma'} \times [0,1].$$

Consequently,

$$\cup_{s\in[0,1]}\zeta_s^{-1}(0) = \cup_{s\in[0,1]}\varphi(Fix\ (g_s)) = \cup_{s\in[0,1]}\varphi(Fix\ (r \circ \widetilde{g}_s))$$

is contained in the *compact subspace* $\varphi(\overline{T}_{\gamma'}^{n,1-\gamma'}) \times [0,1]$ of $\mathbb{R}_+^n \times [0,1]$, and we have achieved our aim.

5.2 Evolutions Based on a Model With Weakened Boundary Assumptions

(i) It might be argued that the presented model of an exchange economy without Walras' law and homogeneity from the preceding Section 5.1 is still not quite satisfactory from the viewpoint of economic intuition on account of its boundary assumptions. Actually, one may desire that excess demand need not be finite for all price vectors from the boundary. Rather, it should also be admitted that excess demand for a commodity may reach infinity when its price tends to zero. Furthermore, it might be argued that the somewhat artificial looking, though economically reasonable index subset $J(p^m)$ from Property (1) should be replaced by the more immediate set $I(p^m) := \{i \in \{1,\ldots,n\}|(p_i^m)$ is unbounded$\}$.

The main question is clearly whether it will be still possible to provide an equilibrium equivalent self–mapping if the basic model from Section 5.1 is modified in this way. Actually, we will be able to show that at the cost of some additional technical effort the constructions from the preceding Section can suitably be adapted. Let us take these points one at a time.

First let us state the complete definition of a *basic exchange economy without Walras' law and homogeneity of type II*. We will present it in the most general form where each market excess demand function may have an individual "exception subset" where it is not defined.

Thus, let us first choose for each $i \in \{1, \ldots, n\}$ an arbitrary *closed* subset D_i of the coordinate hyperplane $\{p \in \partial\mathbb{R}^n_+ | p_i = 0\}$ of $\mathbb{R}^n_+$. Furthermore, for economic reasons the origin 0 shall be contained in every set D_i. A *basic economy* without Walras' law and homogeneity of type II is now given by n continuous market excess demand functions

$$\zeta_i : \mathbb{R}^n_+ \backslash D_i \longrightarrow \mathbb{R}, \quad i = 1, \ldots, n,$$

with the properties:

(1) Let (p^m) be an arbitrary unbounded sequence in $\mathbb{R}^n_+$. Define $I(p^m) := \{i \in \{1, \ldots, n\} | (p_i^m) \text{ is unbounded}\}$. Then there is an index $\tilde{i} \in I(p^m)$ and a natural number $k_o \in \mathbb{N}$ such that

$$\zeta_{\tilde{i}}(p^k) < 0$$

for all $k \geq k_o$.

(2) Let y be an arbitrary point of $\partial\mathbb{R}^n_+ \backslash (\cup_{i=1}^n D_i)$. Then at least for one index $\tilde{j} \in \{1, \ldots, n\}$ with $y_{\tilde{j}} = 0$ one has

$$\zeta_{\tilde{j}}(y) > 0.$$

If (z^m) is a sequence in $\mathbb{R}^n_+ \backslash D_i$ which converges to some $z \in D_i$, then

$$lim_m \zeta_i(z^m) = +\infty.$$

(3) There is a positive real number b such that for all $i = 1, \ldots, n$ and any $p \in \mathbb{R}^n_+ \backslash D_i$ one has

$$\zeta_i(p) > -b.$$

Let us briefly comment on this definition. Actually, it generates a set of spaces of economies parametrized with n-tuples $(D_1, \ldots, D_n)$ of exception subsets. To be sure it is well possible to endow this parametrized set of spaces of economies with a topology, namely with the topology which is derived from the Hausdorff–distance between any two admissible sets D_{i_1} and $D_{i_2}, i = 1, \ldots, n$. However, we will not pursue this further since we will not need this topology in our study.

Note particularly that there is no trouble with admitting (p^m) to lie in the whole $\mathbb{R}^n_+$ in Property (1) since any exception subset D_i is contained in $\{p \in \partial\mathbb{R}^n_+ | p_i = 0\}$.

Corresponding to the n market specific domains of a basic economy one needs n one–parametrizations, i.e. n market evolutions, to define an evolution of economies in the present basic set–up:

Definition 5.3. *An evolution of economies in the presented basic model without Walras' law and homogeneity and with weakened boundary assumptions, or an* **exchange–IV–evolution** *for short, is given by n C^o–uniformly continuous market evolutions, i.e. n market evolutions*

$$(\zeta_{i_s})_{s\in[0,1]} : (\mathbb{R}^n_+ \backslash D_i) \times [0,1] \longrightarrow \mathbb{R}, \quad i = 1,\ldots,n.$$

Analogously to the preceding basic model from Section 5.1 again we can replace the C^o–uniform continuity assumption by the following *uniformized boundary condition*: for any sequence (p^m, s^m) in $\mathbb{R}^n_+ \times [0,1]$ where (p^m) is unbounded and (s^m) converges to some $s^o \in [0,1]$ there is at least one index $\widetilde{i} \in I(p^m)$ and one subsequence (p^{m_k}, s^{m_k}) such that for all k

$$\zeta_{\widetilde{i}_{s^{m_k}}}(p^{m_k}) < 0.$$

We are still left here to ensure that the considerations in Subsection 5.1 (i) establishing the essential differences to the Walrasian exchange model also apply to the present model. Nevertheless, this is immediate. Moreover, in the present set–up it is even necessary to impose further bounding restrictions on each market excess demand function ζ_i in a neighborhood of the exception subset D_i. Otherwise, the additional $n + 1$. "buffer" market $\overline{\zeta}_{n+1}(p, 1) := -\sum_{i=1}^n [(\zeta_i \circ \varphi)(p)] \cdot p_i$ need not be bounded from below. In other words, the differences to the traditional Walrasian model are still increased.

(ii) In the introductory remarks to this Chapter we have already indicated that the weakening of the boundary assumptions involves some additional technical effort with the construction of the equilibrium equivalent self–mapping. Particularly, this applies to the introduction of the more intuitive index subset $I(p^m)$.

On the other hand, the generalization of the price domain easily fits into the constructions provided in Section 5.1 (ii). However, there is one point where we have to be cautious. We must replace the term $(\zeta_{s_h} \circ \varphi)(p)$ in the definition of $\widetilde{g}_{s_h}$ by the term $min\ [(\zeta_{s_h} \circ \varphi)(p); \alpha]$ with some arbitrary real number α greater than $+1$. This is necessary because otherwise the mapping $\widetilde{g}_s$ need not be continuous. Actually, the first summand

$$\lambda(p) \cdot [(\zeta_{s_h} \circ \varphi)(p) + p_h] + (1 - \lambda(p))\frac{1}{10n}$$

of $\widetilde{g}_{s_h}$ need not converge to zero for a sequence (p^m) in $\overline{T}^n$ whose image sequence $(\varphi(p^m))$ converges to some point from D_h.

Now let us examine which adaptions of the constructions from Section 5.1 (ii) are necessary through the introduction of the simplified index subset $I(p^m)$. Actually, the only thing we must adapt is the homeomorphism

$$\varphi : T^n \backslash \Delta^{n-1} \xrightarrow{\approx} \mathbb{R}^n_+.$$

What we need instead is a geometrically intuitive homeomorphism χ which is not only simplex–preserving, but also "collar–preserving". The latter means that the "$\frac{1}{10n}$–collar" of T^n, i.e. the $\frac{1}{10n}$–neighborhood of the subspace of the boundary $\partial T^n \backslash \Delta^{n-1}$ in T^n, is in a simple way mapped into the $\frac{1}{10n}$–collar of $\mathbb{R}^n_+$, i.e. into the $\frac{1}{10n}$–neighborhood of the boundary of $\mathbb{R}^n_+$ in $\mathbb{R}^n_+$. Together with the simplex–preserving property this means that the χ–image of the $\frac{1}{10n}$–collar of any simplex, $0 < t < 1$, is just the $\frac{1}{10n}$–collar of the χ–image of $\Delta^{n-1,t}$. Clearly, the $\frac{1}{10n}$–collar of $\Delta^{n-1,t}$ is the intersection of $\Delta^{n-1,t}$ with the $\frac{1}{10n}$–collar of $\mathbb{R}^n_+$. Figure 5.3 gives an illustration.

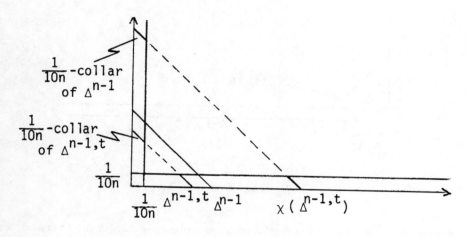

Fig. 5.3: Boundary Condition Areas III

Writing ϵ instead of $\frac{1}{10n}$ the basic idea for constructing a geometrically intuitive homeomorphism

$$\chi : \overline{T}^n \backslash \Delta^{n-1} \longrightarrow \mathbb{R}^n_+$$

with the desired characteristics is to take for χ the identity mapping on $T^{n,n\epsilon} := \cup_{0 < t \leq n\epsilon} \Delta^{n-1,t} \cup \{0\}$ and to extend it over the remaining area $(\overline{T}^n \backslash \Delta^{n-1}) \backslash T^{n,n\epsilon}$ of the domain in the following way: any t–simplex $\Delta^{n-1,t}$ with $n\epsilon < t < 1$ is parallelly stretched outwards onto the simplex $\Delta^{n-1,\sigma(t)}$, where

$$\sigma :]n\epsilon, 1[\longrightarrow]n\epsilon, \infty[$$

is an arbitrary continuous, onto, and strictly monotonically increasing map. Thus the ϵ–collar of $\Delta^{n-1,t}$ is canonically one–to–one mapped onto the ϵ–collar of the image simplex $\Delta^{n-1,\sigma(t)}$. The *inner part of* $\Delta^{n-1,t}$, $\{y \in \Delta^{n-1,t} | \forall\, i = 1, \ldots, n : y_i > \epsilon\}$, is stretched by central projection from the center $(\epsilon, \ldots, \epsilon)$. For $n = 2$ this means that the ϵ–collar of $\Delta^{n-1,t}$ is just parallelly shifted outwards (see Figure 5.3).

In higher dimensions the ϵ–collar is not only shifted, but also stretched, but in a very canonical way (imagine $n = 2$, or 3). The following Figure 5.4 illustrates the collar–preserving homeomorphism χ in the case $n = 2$. It still

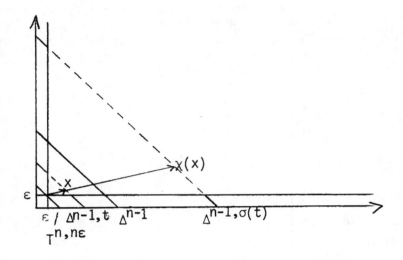

Fig. 5.4: Boundary Condition Areas IV

remains to verify that χ is indeed a homeomorphism. An appropriate criterion for this is the following standard result from general topology.

Proposition 5.4. *A map $f : X \longrightarrow Y$ between topological spaces X and Y is a homeomorphism if and only if f is (1) continuous, (2) f is one–to–one and onto, and (3) f^{-1} is also continuous.*

The geometrical description of χ immediately reveals the three required properties. However, it is a natural question how this intuitive, geometrical representation of χ and its inverse can be translated into exact analytical terms.

To spare room we just present the solution to this question (the reader interested in a detailed construction which is further supported by geometrical heuristics is referred to Lehmann–Waffenschmidt (1987c, Section 3). For notational convenience we abbreviate $\sum_{i=1}^{n} x_i$ by $\sum_x$. For our task the following intuitive "cutting" mapping will turn out to be of major value

$$\alpha : (\overline{T}^n \backslash \Delta^{n-1}) \backslash \overline{T}^{n-1,n\epsilon} \longrightarrow \left[0, \frac{1}{10n}\right]$$

$$x \mapsto min_{i=1,\ldots,n} \left(x_i; \frac{1}{10n}\right).$$

Now, χ is given by

$$\chi \; : \; \bar{T}^n \backslash \Delta^{n-1} \overset{\approx}{\to} \mathbb{R}^n_+$$

$$x \mapsto \begin{cases} x, & \text{for } x \in \bar{T}^{n,n\epsilon} \\[2ex] \begin{pmatrix} a(x) \\ \vdots \\ a(x) \end{pmatrix} + \dfrac{n\epsilon\left(\frac{1-n\epsilon}{1-\sum_x}\right) - na(x)}{\sum_x - na(x)} \begin{pmatrix} x_1 - a(x) \\ \vdots \\ x_n - a(x) \end{pmatrix}, & \text{otherwise} \end{cases}$$

Its inverse χ^{-1} is given by

$$\chi^{-1} \; : \; \mathbb{R}^n_+ \longrightarrow \bar{T}^n \backslash \Delta^{n-1}$$

$$y \mapsto \begin{cases} y, & \text{for } y \in \bar{T}^{n,n\epsilon} \\[2ex] \begin{pmatrix} a(y) \\ \vdots \\ a(y) \end{pmatrix} + \dfrac{\left[\left(1 - n\epsilon\frac{1-n\epsilon}{\sum_y}\right) - na(y)\right]}{\sum_y - na(y)} \begin{pmatrix} y_1 - a(y) \\ \vdots \\ y_n - a(y) \end{pmatrix}, & \text{otherwise.} \end{cases}$$

Verifications are not hard. They are made explicit in the step–by–step construction in Lehmann–Waffenschmidt (1987c, Section 3).

For our concluding considerations, however, the *intuitive geometrical description* of χ is the more convenient one. Going through the constructions from Section 5.1 (ii) it is evident that everything equally works for a basic economy in our present context where φ is replaced by χ. Actually, the proof of the preparatory result (1) is even shorter. From the definitions it follows immediately that $I(\chi(y^{m_k})) \subseteq H(y^{m_k})$. To be accurate the formulation of the technical result (2) must be changed into "... $\zeta_{s_{i_y}}(\chi(z)) > \delta$ whenever $\chi(z) \notin D_{i_y}$." Accordingly, the corresponding result (2') for an exchange–III–evolution must also be changed.

(iii) It remains to ensure that an exchange–IV–evolution induces a continuous one-parametrization of equilibrium equivalent self–mappings. But this is completely straightforward from the preceding considerations.

6

Evolutions in a General Equilibrium Framework With Production, Taxes, and Subsidies

In this Chapter we are going to present a notably comprehensive static framework of general equilibrium which includes production and also a wide variety of tax and subsidy schemes. We employ the formulation by Kehoe (1985b). He ensures overall consistency of the model by providing a proof of existence of equilibrium that is both simpler and more general than those given previously by, for example, Shoven and Whalley (1973) and Todd (1979). Kehoe considers this framework particularly intended for researchers who employ empirical general equilibrium models for policy analysis (cf. Kehoe, 1985b, p. 315). Actually, for the purposes of our study this framework has still a further remarkable advantage: it contains parameters which in principle can be controlled by an economic policy institution. This will become important for our later applications in Chapter 19.

Kehoe's first model which we will present in Section 6.1 below allows for production and taxes, whereas his second model (see Section 6.2 below) also includes subsidies.

An equilibrium in these models means a state of the endogenous variables which simultaneously makes all interdependent plans by the government and the individuals consistent. Thus, from the equilibrium theoretical viewpoint there is no circularity problem with the mutually dependent tax redisbursals and tax payments.

We will present both of Kehoe's model versions following our usual agenda. As in the preceding Chapter for the sake of easier understanding we first will carry out all necessary verifications for the first version of the model and then point out the extensions needed for the generalized second one. According to our principles we carefully fit our notations into Kehoe's.

6.1 Evolutions Based on a General Equilibrium Model With Production and Taxes

(i) In Kehoe's first version of an equilibrium model with production and taxes (1985b, pp. 318–321) an agent can be a consumer, a producer, or the government. Aggregate excess demand on the n commodity markets is given by a C^1 function

$$\zeta : \mathbb{R}^n_+ \backslash \{0^n\} \times \mathbb{R}_+ \longrightarrow \mathbb{R}^n$$
$$(p, r) \longmapsto \zeta(p, r).$$

r stands for the *total tax revenue*. ζ is homogeneous of degree zero, bounded from below, and satisfies the following intuitive boundary assumption

$$*)\quad k_\zeta : \mathbb{R}_+ \longrightarrow \mathbb{R}_+$$
$$r \longmapsto \inf_{p \in \mathbb{R}^n_+ \backslash \{0^n\}} \{||\zeta(p, r)||\}$$

$$\text{satisfies } \lim_{r \longrightarrow \infty} k_\zeta(r) = \infty.$$

In words the boundary assumption*) just means that with the r-argument growing beyond all finite bounds the absolute value of the excess demand function $||\zeta(p, r)||$ also grows beyond all finite bounds *all over the price space* $\mathbb{R}^n_+ \backslash \{0^n\}$. This property of the excess demand function particularly ensures the intuitive requirement that for any real $\alpha > 0$ there is a real $\beta > 0$ such that for all $p \in \mathbb{R}^n_+ \backslash \{0\}$ one has: $r > \beta \Rightarrow ||\zeta(p, r)|| > \alpha$. Later we will see that this implies that the equilibrium set in fact is contained in a compactum which is essential for the construction of an equilibrium equivalent self-mapping (see Kehoe, 1985b, p. 321 first paragraph). Unfortunately, Kehoe's original condition

$$\text{given any } p \in \mathbb{R}^n_+ \backslash \{0^n\}$$

$$\lim_{r \longrightarrow \infty} ||\zeta(p, r)|| = \infty \tag{6.1}$$

is too weak to ensure that. (Actually, it is not hard to find counterexamples.) However, the following additional requirement to (6.1) obviously makes it sufficient for the purposes of the later analysis: the family of *partial functions*

$$\zeta(p, -) : \mathbb{R}_+ \longrightarrow \mathbb{R}^n$$

which is parametrized by the admissible price vectors $p \in \mathbb{R}^n_+ \backslash \{0^n\}$, is C^0-uniformly convergent on the whole domain $\mathbb{R}_+$ for varying p.

The tax payments generated by consumption and income taxation are specified by a C^1 function

$$t : \mathbb{R}^n_+ \backslash \{0^n\} \times \mathbb{R}_+ \longrightarrow \mathbb{R}_+$$
$$(p, r) \longmapsto t(p, r)$$

which is homogeneous of degree one, i.e. $t(\lambda p, \lambda r) = \lambda t(p, r)$ for $\lambda > 0$. Tax payments $t(p, r)$ and the tax revenue r are expressed in the same units of account as expenditures $\zeta(p, r)p$. The function t furthermore satisfies an appropriately modified version of Walras' law, namely the aggregate budget constraint

$$\zeta(p, r)p + t(p, r) = r$$

for all admissible (p, r). Note that this aggregate budget constraint still does not indicate how the total tax revenues r is actually generated. Generation of r will be become clear below when also the production sphere will be introduced.

To enhance the reader's intuition on this specification let us have a look on Kehoe's *example* (1985b, pp. 318–319) of an economy with h consumers. At prices p the j-th consumer's income is given by the value of his initial endowment bundle

$$\sum_{i=1}^{h} p_i \omega_i^j$$

plus his share of tax revenue,

$$\theta_j r.$$

Clearly, the vector of share coefficients $(\theta_1, \ldots, \theta_h)$ lies in Δ^{h-1}. For instance, $\theta_1 = \ldots = \theta_{h-1} = 0$ and $\theta_h = 1$, where agent h is the government. The endowment income $\sum_{i=1}^{h} \rho_i \omega_i^j$ of each consumer j is taxed at a rate $\rho_j \in [0, 1[$, and consumer j's final demand for commodity i is taxed 'ad valorem' at a rate $\tau_{ij} \in [0, 1[$ on its value. Accordingly, the utility maximization problem of consumer j is the following:

$$\text{max } u_j(x_1^j, \ldots, x_n^j)$$
$$\text{so that } \sum_{i=1}^{h} p_i(1 + \tau_{ij})x_i^j \le (1 - \rho_j) \sum_{i=1}^{h} \rho_i \omega_i^j + \theta_j r$$
$$x_i^j \ge 0 \text{ for all } i, j.$$

u_j is a strictly concave and monotonically increasing utility function. Thus, agent j's derived excess demand function

$$\zeta^j : \mathbb{R}_+^n \backslash \{0^n\} \longrightarrow \mathbb{R}_+ \longrightarrow \mathbb{R}^n$$
$$(p, r) \mapsto \begin{pmatrix} x_1^j(p, r) - \omega_1^j \\ \vdots \\ x_n^j(p, r) - \omega_n^j \end{pmatrix}$$

is continuous and the aggregate excess demand function $\sum_{j=1}^{h} \zeta^j$ satisfies for any $p \in \mathbb{R}_+^n \backslash \{0^n\}$ the condition $\|\zeta(p, r^m)\| \longrightarrow +\infty$ as $r^m \longrightarrow +\infty$. This condition means that, anything else being equal, if tax revenue becomes arbitrarily large, then the income of at least one consumer (the government for instance) becomes arbitrarily large, which in turn implies that excess demand for some good becomes arbitrarily large (cf. Kehoe, 1985b, p. 319). Accordingly, the tax payment function t is specified by

$$t(p,r) = \sum_{j=1}^{r} \rho_j \left(\sum_{i=1}^{n} p_i \omega_i^j \right) + \sum_{j=1}^{h} \left(\sum_{i=1}^{n} \tau_{ij} p_i x_i^j(p,r) \right).$$

t is C^1 and homogeneous of degree one as long as the x_i^j are. Since each individual demand function satisfies the budget constraint with equality, ζ and t satisfy the modified Walras' law. (For a further example which also allows for tax rates and revenue shares varying with income the reader is referred to Kehoe (1985b, p. 319, last paragraph).

Now let us come back again to the general model. The production sphere is specified by an $n \times m$ activity analysis matrix $A = (a_{ij})$ with the following properties:

(1) A induces n free disposal activities, one for each commodity. Formally, this means that the last n columns of A form the negative n-dimensional unit-matrix.
(2) There is no output without inputs, i.e.

$$\{x \in \mathbb{R}^n | x = Ay, \ y \geq 0^n\} \cap \mathbb{R}_+^n = \{0^n\}.$$

Production taxes are specified by an $n \times m$ matrix $A^* = (a_{ij}^*)$ with $a_{ij}^* = a_{ij} - \sigma_{ij} |a_{ij}|$ where $\sigma_{ij} \in [0, 1]$. This means, input or output of commodity i in activity j is taxed at a rate of $\sigma_{ij} \in [0, 1]$. Thus,

(3) $-2|A| \leq A^* \leq A$

Furthermore, there are no taxes at free disposal activities. Accordingly, the *revenue generated* by production taxes at prices p and at activity levels $y \in \mathbb{R}_+^m$ is

$$p'(A - A^*)y \geq 0.$$

An *economy* is now defined as a quadruple (ζ, t, A, A^*). An *equilibrium* of an economy is a pair $(p^0, r^0) \in \Delta^{n-1} \times \mathbb{R}_+$ of a price system p^0 and a tax revenue amount r^0 that satisfies the following conditions,

(E.1) $p^{0'} A^* \leq 0^m$.
(E.2) there is a $y^0 \in \mathbb{R}^m \backslash \{0^m\}$ such that $\zeta(p^0, r^0) = Ay^0$.
(E.3) $r^0 = t(p^0, r^0) + p^{0'}(A - A^*)y^0$

Let us briefly comment on these equilibrium conditions. From (E.2) and Walras' law follows immediately (E.3) $\Leftrightarrow p^{0'} A^* y^0 = 0$. Actually, it is this equivalent formulation of equilibrium condition (E.3) which we will use in our later constructions.

Together with (E.1) the alternative formulation of (E.3) implies that after-tax profits are maximized at an equilibrium.

(E.2) means that excess demand actually can be supplied by the producers. (E.3) expresses the fact that in equilibrium the redisbursals of the total tax revenue equal the total tax receipts $t(p^0, r^0) + p^{0'}(A - A^*) y^0$. The normalization expressed by $p^0 \in \Delta^{n-1}$ is obviously permitted by the homogeneity properties of ζ and t.

The formalization of an *evolution of economies with production and taxes* is straightforward:

Definition 6.1. *An **evolution of economies with production and taxes** is a quadruple of four continuous one-parametrizations $(\zeta_s, t_s, (a_{ij_s}), (a_{ij_s}^*))_{s \in [0,1]}$ such that, moreover, the two component one-parametrizations $(\zeta_s)_{s \in [0,1]}$ and $(t_s)_{s \in [0,1]}$ are C^0-uniformly continuous and the one-parametrization $(a_{ij_s})_{s \in [0,1]} = (A_s)_{s \in [0,1]}$ satisfies the condition that for any $w \in \mathbb{R}_+^n$ with $\zeta_s(-, -) \geq -w$ for all $s \in [0,1]$ on the whole domain $(\mathbb{R}_+^n \backslash \{0\}) \times \mathbb{R}_+$ the set*

$$\{x \in \mathbb{R}^n | \exists y \geq 0, \ \exists s \in [0,1] : x = A_s(y) \ and \ x \geq -w\}$$

is bounded.

Note that the last assumption is just a uniformization of the assumption (2) 'no output without inputs' on the production matrix of static economies. This is formalized by the second equivalence of the following chain of equivalences.

Proposition 6.2. $\forall_{w \in \mathbb{R}_+^n} \{x \in \mathbb{R}^n | x \in \bigcup_{s \in [0,1]} A_s(\mathbb{R}_+^m) \ and \ x \geq -w\}$ *is bounded, i.e. there is an $\alpha_w > 0$ such that $||x|| < \alpha_w$ for all x from this set*

$\Leftrightarrow \{x \in \mathbb{R}^n | x \in \bigcup_{s \in o,1]} A_s(\mathbb{R}_+^m) \ and \ x \geq (-1, \ldots, -1)\}$ *is bounded*

$\Leftrightarrow \overline{\bigcup_{s \in [0,1]} A_s(\mathbb{R}_+^m)} \cap \mathbb{R}_+^n = \{0^n\}$

$\Leftrightarrow$ *there is a closed subspace $K \subset \mathbb{R}^n$ with $\overline{\bigcup_{s \in [0,1]} A_s(\mathbb{R}_+^m)} \subset K$ and $K \cap \mathbb{R}_+^n = \{0^n\}$.*

Proof. The first and the last equivalence are trivial, whereas the crucial middle one is not.

"$\Rightarrow$:" Let us abbreviate $N := \bigcup_{s \in [0,1]} A_s(\mathbb{R}_+^m)$ and assume that there is an $x \in \overline{N} \cap \mathbb{R}_+^n$ with $x \neq 0$. Then there is a sequence x^k in N with $x^k \longrightarrow x$. Without loss of generality we may assume that $||x^k|| = ||x|| = 1$ and $x_i^k \geq -1$ for all k and $1 \leq i \leq n$. Define

$$m^k := \max\{|x_i^k| | x_i^k < 0\}.$$

Note that the right set is non-empty since $N \cap \mathbb{R}_+^n = \{0^n\}$. The latter is due to the assumption 'no output without inputs'. Since $x^k \longrightarrow x \in \mathbb{R}_+^n$, the sequence m^k converges to zero. Put $y^k := \frac{x^k}{m^k}$. Clearly $y^k \in N$ and $||y^k|| \longrightarrow \infty$ for $k \longrightarrow \infty$. If we can show that $y^k \geq (-1, \ldots, -1)$ for all k, then the presumption that $\{x \in N | x \geq (-1, \ldots, -1)\}$ is bounded contradicts $||y^k|| \longrightarrow \infty$. Consequently, the assumption that there is an $x \in \overline{N} \cap \mathbb{R}_+^n$ with $x \neq 0$ is wrong. Now let us choose any $j \in \{1, \ldots, n\}$. Clearly, $y_j^k < 0$. From $|x_j^k| \leq m^k$ follows $|y_j^k| = \frac{|x_j^k|}{m^k} \leq 1$. But this means that $y_j^k \geq -1$, and we are done.

"$\Leftarrow$": We begin with the observation that $K := \overline{N} \cap S^{n-1}$ is compact. Define

$$\lambda : K \longrightarrow \mathbb{R}_+$$
$$x \mapsto \min_{\substack{i \text{ with} \\ x_i < 0}} \frac{1}{|x_i|} = \frac{1}{\left\lceil \max_{\substack{i \text{ with} \\ x_i < 0}} |x_i| \right\rceil}.$$

Actually, λ is well-defined since $K \cap \mathbb{R}_+^n = \emptyset$ by presumption. Moreover, λ is continuous. Consequently, there is a $\lambda_0 > 0$ with

$$\forall_{x \in K} \ \lambda(x) \leq \lambda_0.$$

Choose now any $x \in \overline{N}$ with $0^n \neq x$ and $x \geq (-1, \ldots, -1)$. Define $z := \frac{x}{||x||} \in K$. There is an $\tilde{i} \in \{1, \ldots, n\}$ such that $z_{\tilde{i}} < 0$ and $\lambda(z) = \frac{1}{|z_{\tilde{i}}|}$. Clearly, $\frac{1}{|z_{\tilde{i}}|} = \frac{||x||}{|x_{\tilde{i}}|} \leq \lambda_0$, and from $-1 \leq x_{\tilde{i}} \leq 0$ follows

$$||x|| \leq \lambda_0 |x_{\tilde{i}}| \leq \lambda_0.$$

This means that any $x \in \overline{N}$ with $x \geq (-1, \ldots, -1)$ lies in the n-ball $\overset{\circ}{B}_{\lambda_0}^n(0^n)$, and this completes the proof. $\qquad \square$

(ii) Now, we are going to provide an equilibrium equivalent self-mapping for the presented model. Before reporting on Kehoe's construction we have to do a last preparatory step (cf. Kehoe (1985b), p. 321, first paragraph). We have to ensure that in equilibrium tax revenues cannot exceed some fixed upper bound $\beta > 0$. This implies that all candidates for equilibria lie in the compact convex set $\Delta^{n-1} \times [0, \beta]$ which will be crucial for our constructions. The existence of such a β can be seen in the following way: the boundedness from below of ζ, say by $-w$, $w \in \mathbb{R}_+^n$, and assumption (2) on the production sphere clearly imply that the *production possiblity set* $P := \{x \in \mathbb{R}^n | x \geq -w, \ x = Ay$ for some $y \geq 0^m\}$ is *bounded*, i.e. there is a real $\alpha > 0$ such that $P \subset \overset{\circ}{B}_\alpha^n$. Furthermore, due to boundary assumption* there is clearly a real $\beta > 0$ so that $||\zeta(p, r)|| \geq \alpha$ for any pair $(p, r) \in \Delta^{n-1} \times [\beta, \infty[$. But this implies that all equilibria already must lie in $\Delta^{n-1} \times [0, \beta]$.

Kehoe proposes the following construction for an equilibrium equivalent self-mapping (1985b, pp. 321–322):

$$g : \Delta^{n-1} \times [0, \beta] \longrightarrow \Delta^{n-1} \times [0, \beta]$$
$$(p, r) \longmapsto (x, y)$$

where (x, y) solves the following program:
min $1/2[(x - p - \zeta(p, r))(x - p - \zeta(p, r)) + (y - t(p, r))^2]$
so that

(1) $(x, y) \in \Delta^{n-1} \times [0, \beta]$
(2) $x'A - (1 + y - r)p'(A - A^*) \leq 0^m$.

We have to verify four issues:

(1) The constraint set is non-empty.

This follows directly from assumption (2) on the production sphere.

(2) The constraint set is a subset of $\Delta^{n-1} \times [0, \beta]$.

This follows from the assumption that there are no taxes on free disposal activities.

(3) $g(p, r)$ is continuous.

This follows from the facts that for any pair of arguments (p, r) the constraint set obviously is closed and convex and varies continuously as a point-to-set mapping, and the objective function of the program is strictly convex. The latter follows from the positive definiteness of the Hesse matrix of the objective function (recall that (p, r) is fixed): its gradient is

$$1/2 \begin{pmatrix} 2x_1 - 2(p_1 + \zeta_1(p, r)) \\ \vdots \\ 2x_n - 2(p_n + \zeta_n(p, r)) \\ 2y - 2t(p, r) \end{pmatrix},$$

and consequently the $(n + 1) \times (n + 1)$ Hesse matrix becomes

$$1/2 \begin{pmatrix} 2 & & 0 \\ & \ddots & \\ 0 & & 2 \end{pmatrix} = \begin{pmatrix} 1 & & 0 \\ & \ddots & \\ 0 & & 1 \end{pmatrix}.$$

4) (p^0, r^0) is an equilibrium of (ζ, t, A, A^*) if and only if it is a fixed point of the associated mapping g.

This is shown in the proof of Theorem 1 by Kehoe (1985b, p. 322).

(iii) From the assumptions on an evolution of economies in the present context, from the definition of the equilibrium equivalent self-mapping g, and from the considerations in (3) above it follows directly that any admissible evolution of economies with production and taxes induces a *continuous* one-parametrization of equilibrium equivalent self-mappings, as desired.

6.2 Evolutions Based on a General Equilibrium Model With Production, Taxes, and Subsidies

(i) In his paper (1985b) Kehoe finally extends the model with production and tax schemes we have dealt with to a model also allowing for a wide variety of subsidy schemes (Kehoe, ibid., pp. 329–331).

> "We allow the same sorts of subsidies as we do taxes: ad valorem and specific subsidies on production and consumption and linear and non-linear subsidies on income. We need to be able to guarantee, however, that the government can pay these subsidies out of its tax revenues. To do this, we introduce another variable γ, that is equal to the fraction of the subsidy payments the government can afford to make: If $\gamma = 1$, the government has enough tax revenues to make all the subsidy payments. If $\gamma = 0$, the government cannot afford to make any subsidy payments." (Kehoe, 1985b, pp. 329–330)

Analogously to the first model version aggregate excess demand is given by a C^1 function

$$\zeta : \mathbb{R}^n_+ \setminus \{0^n\} \times \mathbb{R}_+ \times [0,1] \longrightarrow \mathbb{R}^n$$
$$(p, r, \gamma) \longmapsto \zeta(p, r, \gamma),$$

and tax payments by consumers net of subsidies to consumers are specified by a C^1 function

$$t : \mathbb{R}^n_+ \setminus \{0^n\} \times \mathbb{R}_+ \times [0,1] \longrightarrow \mathbb{R}_+$$
$$(p, r, \gamma) \longmapsto t(p, r, \gamma).$$

For any fixed $\gamma \in [0,1]$, ζ and t are assumed to satisfy the same assumptions as for the first model. Walras' law is now to be read as

$$\zeta(p, r, \gamma)p + t(p, r, \gamma) = r.$$

Furthermore, it is assumed that

$$t(p, r, 0) > 0$$

for any pair $(p, r) \in (\mathbb{R}^n_+ \setminus \{0^n\}) \times \mathbb{R}_+$.

Subsidies to producers are modelled by an $(n \times m)$-matrix $A^{**} = (a^{**}_{ij}) = (\chi_{ij}|a_{ij}|)$ where $\chi_{ij} \geq 0$ is the ad valorem subsidy rate on the output or the input of commodity i in activity j. There are no subsidies on the n free disposal activities, that means $\chi_{ij} = 0$ for $i = m - n + 1, \ldots, m$, and $j = 1, \ldots, n$.

Extending the example from the preceding Section to our new situation consumer j's problem becomes the following maximization program

$$\max \ u_j(x^j_1, \ldots, x^j_n)$$

so that

$$\sum_{i=1}^{n} p_i(1 + \tau_{ij} - \gamma\sigma_{ij})x_i^j \leq (1 - \rho_j)\sum_{i=1}^{n} p_i\omega_i^j - \gamma q(p)\eta_j + \theta_j r.$$

$\sigma_{ij} \in [0,1[$ is the ad valorem subsidy rate on consumer j's final demand for commodity i, $\eta_j \geq 0$ is some fixed income transfer to consumer j, and $q(p)$ is some price index which is homogeneous of degree one.

To sum up an *economy* is given by a quintuple $(\zeta, t, A, A^*, A^{**})$. An *equilibrium* is a triple $(p^0, r^0, \gamma^0) \in \Delta^{n-1} \times \mathbb{R}_+ \times [0,1]$ of a price system p^0, a tax revenue amount r^0, and a subsidies realization rate γ^0 which satisfies the reasonable properties

(E'.1) $p^{0'}(A^* + \gamma^0 A^{**}) \leq 0^m$
(E'.2) $\zeta(p^0, r^0, \gamma^0) = Ay^0$ for some $y \in \mathbb{R}_+^m$
(E'.3) $r^0 = t(p^0, r^0, \gamma^0) + p^{0'}(A - A^* - \gamma^0 A^{**})y^0$
(E'.4) If $\gamma^0 < 1$, then $r^0 = 0$.

Thus equilibria are from the subspace $\{(p, 0, s)\} \cup \{(p, r, 1)\}$ of $\Delta^{n-1} \times \mathbb{R}_+ \times [0,1]$. Analogously to the preceding model (E'.3) can be equivalently reformulated:

$$(E'.3) \Leftrightarrow p^{0'}(A^* + \gamma^0 A^{**})y^0 = 0.$$

Again, it will be this reformulated form which will be crucial for our later constructions.

The new condition (E'.4) says that in equilibrium the first committment of the government's tax receipts is to subsidies, and only after making all the subsidy payments can it transfer revenue to consumers. Furthermore, (E'.3) and (E'.4) together imply that if $\gamma^0 = 0$ then $t(p^0, r^0, 0) + \pi^{0'}(A - A^*)y^0 = 0$ (Kehoe, 1985b, p. 330).

Definition 6.3. *An* **evolution of economies with production, taxes, and subsidies** *is in complete analogy to the preceding model in Section 3.1 given by a quintuple of C^0-uniformly continuous one-parametrizations $(\zeta_s, t_s, (a_{ij_s}), (a_{ij_s}^*), (a_{ij_s}^{**}))_{s \in [0,1]}$. Again the uniformized boundary condition is required:*
for any sequence (p, r^m, γ, s^m) in $(\mathbb{R}_+^n \setminus \{0^n\}) \times \mathbb{R}_+ \times [0,1]^2$ with $s^m \longrightarrow s^0$ and $r^m \longrightarrow +\infty$ one has $||\zeta_{s^m}(p, r^m, \gamma)|| \longrightarrow +\infty$.

(ii) The following extension of the mapping g from the preceding Section serves as equilibrium equivalent self-mapping for the present model version (see Kehoe, 1985b, pp. 330–331):

$$g' : \Delta^{n-1} \times [0, \beta] \times [0,1] \longrightarrow \Delta^{n-1} \times [0, \beta] \times [0,1]$$
$$(p, r, \gamma) \mapsto (x, y, v)$$

where (x, y, v) solves the minimization program
$$\min 1/2[(x - p - \zeta(p, r, \gamma)) \cdot (x - p - \zeta(p, r, \gamma)) + (y - t(p, r, \gamma))^2$$
$$+ (r - \gamma - t(p, r, \gamma))^2]$$
so that

(1) $(x, y, v) \in \Delta^{n-1} \times [0, \beta] \times [0, 1]$

(2) $x'A - (1 + y - r + v - \gamma)p'(A - A^* - A^{**}) - (1 + v - \gamma)(1 - \gamma)p'A^{**} \leq 0^m.$

The gradient of the objective function for fixed (p, r, γ) is given by

$$1/2 \begin{pmatrix} 2x_1 - 2(p_1 + \zeta_1(p, r, \gamma)) \\ \vdots \\ 2x_n - 2(p_n + \zeta_n(p, r, \gamma)) \\ 2y - 2t(p, r, \gamma) \\ 2v - 2(\alpha + t(p, r, \gamma)). \end{pmatrix}$$

Thus the $(n + 2) \times (n + 2)$ Hesse matrix is again the identity mapping

$$1/2 \begin{pmatrix} 2 & & 0 \\ & \ddots & \\ 0 & & 2 \end{pmatrix} = \begin{pmatrix} 1 & & 0 \\ & \ddots & \\ 0 & & 1 \end{pmatrix}.$$

The issues (1) to (3) at the end of Subsection 6.1 (ii) immediately carry over to the present situation. The analogue to item (4), i.e. that the set of equilibria of an economy $(\zeta, t, A, A^*, A^{**})$ equals the fixed point set of g', is proven on p. 331 in Kehoe (1985b).

(iii) In complete analogy to Subsection 6.1 (iii) follows that an evolution of economies with production, taxes, and subsidies in fact induces a continuous one-parametrization $(g'_s)_{s \in [0,1]}$ of equilibrium equivalent self-mappings, and we are done.

7

Evolutions in the Temporary Fixprice Equilibrium Framework

We are now going to leave the class of general equilibrium models where all prices are flexible turning to the framework of temporary fixprice equilibrium with quantity constraints. We will present two new models which are suitable for our purposes.

The first model is a modified version of the well-known rationing micromodel with effective demand of the Benassy type and a deterministic rationing scheme. The modifications are intuitive and will be crucial for the construction of an equilibrium equivalent self-mapping for this model.

The second model has not been adapted from the literature. It conceptualizes an economy with quantity rationing under temporarily fixed prices and wages in a *multi-sectoral* set-up. The model combines in an intuitive way the principles of the well-known quantity constrained macromodel with the idea of several interdependent sectors (industries). From the macromodel it inherits the opportunity of a geometrical representation in two dimensions. What it makes particularly appealing to the economist is that the prerequisitory problem of existence of equilibrium reduces to a remarkably simple mathematical situation which furthermore is formalizable by an intuitive function. Moreover, it allows for a natural extension of the well-known quantity tâtonnement process of the macromodel (the reader is referred to Lehmann-Waffenschmidt 1987, Section 4, for details).

The primary motivation for such a model is given by the lack in the literature of a model framework with quantity rationing which allows for analyzing the economywide effects of *sectorspecific* policy measures on employment and production. Thinking for instance of the virulent discussion in the European countries on supportive measures for high-technology firms, or on subsidies for the 'sun-set industries' makes the need for such a theoretical framework apparent. Actually, the traditional micromodels from the quantity constrained framework do not provide a useful device for an analysis of sectorspecific policy measures since there are too many and too complex interdependencies due to the complete disaggregation.

Like the models from the preceding Chapter also the two models of the present Chapter have the advantage of containing explicit parameters which in principle can be controlled by some policy institution. For the microversion presented in Section 7.1 these are prices, wages, and the rationing schemes, and for the multi-sectoral version presented in Section 7.2 only prices and wages.

7.1 Evolutions Based on a Quantity Constrained Micromodel With Effective Demand à la Benassy

(i) We consider an economic system where m economic agents interact on n commodity markets. As usual, demand is given positive sign, and supply negative. Prices on all markets are given at a temporarily fixed level $p = (p_1, \ldots, p_n) \in \mathbb{R}^n_+ \backslash \{0^n\}$. Furthermore, each agent $a \in \{1, \ldots, m\}$ individually perceives upper and lower rationing bounds on each of the n markets for his potential demands and supplies. Let us represent the rationing bounds which are subjectively perceived by agent a for his demand by an n-vector $\overline{z}^a \geq 0^n$ and the perceived bounds for supply by an n-vector $\underline{z}^a \leq 0^n$. Under these circumstances each agent a communicates his individual *planned effective demand/supply signals*

$$\widetilde{z}^a := \widetilde{z}^a(p; \underline{z}^a, \overline{z}^a) \in \mathbb{R}^n$$

of the Benassy type to all markets. Particularly this means that agent a's planned effective demand/supply on the i-th market $\widetilde{z}^a_i(p; \underline{z}^a, \overline{z}^a)$ is not necessarily contained in the interval $[\underline{z}^a_i, \overline{z}^a_i]$. Furthermore, every $\widetilde{z}^a$ is a *continuous* function. For the microeconomic foundation of $\widetilde{z}^a$ by an optimization program we refer to the literature.

Final allocation in case of a non-market-clearing price system p is achieved through a *deterministic rationing scheme*. Formally, the rationing scheme is specified by an m-tuple of composite continuous functions $(F^a)_{a=1,\ldots,m}$, with

$$
F^a : \begin{pmatrix} \underline{z}^1 \\ \vdots \\ \underline{z}^m \\ \overline{z}^1 \\ \vdots \\ \overline{z}^m \end{pmatrix} \overset{\widetilde{z}}{\longmapsto} \begin{pmatrix} \widetilde{z}^1(p; \underline{z}^1, \overline{z}^1) \\ \vdots \\ \widetilde{z}^m(p; \underline{z}^m, \overline{z}^m) \end{pmatrix}
$$

$$
\overset{\widetilde{F}^a}{\longmapsto} \begin{pmatrix} F^a_1(\widetilde{z}^1(p; \underline{z}^1, \overline{z}^1)), \ldots, \widetilde{z}^m(p; \underline{z}^m, \overline{z}^m)), \\ \vdots \\ F^a_n(\widetilde{z}^1(p; \underline{z}^1, \overline{z}^1)), \ldots, \widetilde{z}^m(p; \underline{z}^m, \overline{z}^m)) \end{pmatrix}
$$

where $F_i^a(\widetilde{z}^1(p;\underline{z}^1,\overline{z}^1),\dots,\widetilde{z}^m(p;\underline{z}^m,\overline{z}^m)) \in \mathbb{R}$ denotes agent a's *realized demand/supply on market* i. (Subsequently we will also write $F_i^a(\widetilde{z}^1,\dots,\widetilde{z}^m)$ for short.)

It is reasonable to require the following properties for a rationing scheme $(F^a)_{a=1,\dots,m}$:

given planned effective demand/supply signals $\widetilde{z}^1,\dots,\widetilde{z}^m$ by the agents

(1) all realized transactions F_i^a must be *voluntary*. Formally this means that for every $i \in \{1,\dots,n\}$ and $a \in \{1,\dots,m\}$ the following holds: if $\widetilde{z}_i^a(p;\underline{z}^a,\overline{z}^a) \leq 0$, then for all admissible $(\underline{z}^j,\overline{z}^j), j = 1,\dots,\hat{i},\dots,m$, we have $\widetilde{z}_i^a(p;\underline{z}^a,\overline{z}^a) \leq F_i^a(\widetilde{z}^1(p;\underline{z}^1,\overline{z}^1),\dots,\widetilde{z}^m(p;\underline{z}^m,\overline{z}^m)) \leq 0$, and if $\widetilde{z}_i^a(p;\underline{z}^a,\overline{z}^a) \geq 0$, then for all admissible $(\underline{z}^j,\overline{z}^j), j = 1,\dots,\hat{i},\dots,m$ we have $0 \leq F_i^a(\widetilde{z}^1(p;\underline{z}^1,\overline{z}^1),\dots,\widetilde{z}^m(p;\underline{z}^m,\overline{z}^m)) \leq \widetilde{z}_i^a(p;\underline{z}^a,\overline{z}^a)$.

(2) the allocated transactions always must be *feasible*, i.e. for every $i \in \{1,\dots,n\}$ we have $\sum_{a=1}^m F_i^a(\widetilde{z}^1,\dots,\widetilde{z}^m) = 0$.

We furthermore assume that all perceived rationing bounds $\underline{z}_i^a$ and $\overline{z}_i^a$, and all *realized* demand and supply transactions $F_i^a(\dots)$, $a = 1,\dots,m$, are contained in a compact interval $[-b_i, c_i]$, $i = 1,\dots,n$, with b_i and $c_i > 0$. Denote the cuboid $[-b_i, 0] \times \dots \times [-b_n, 0]$ by G_-, and the cuboid $[0, c_1] \times \dots \times [0, c_n]$ by G_+. (Note that $G_- \times G_+ \subset \mathbb{R}_-^n \times \mathbb{R}_+^n$.) Thus,

$$\widetilde{z}^a : \{p\} \times G_- \times G_+ \longrightarrow \mathbb{R}^n$$

for $a = 1,\dots,m$, and

$$F^a : (G_-)^m \times (G_+)^m \longrightarrow [-b_1, c_1] \times \dots \times [-b_n, c_n].$$

(Naturally, one may view $\widetilde{z}^a$ as a function in $\underline{z}^a$ and $\overline{z}^a$ parametrized with the price system $p \in \mathbb{R}_+^n \setminus \{0^n\}$.)

Clearly, this is not a restrictive assumption. Particularly, all *planned* transactions by the agents $\widetilde{z}_i^a(p;\underline{z}^a,\overline{z}^a)$ *need not be contained* in the interval $[-b_i, c_i]$.

To sum up an *economy* in our present context with temporarily fixed prices, quantity constraints and effective demand à la Benassy is specified by a $(2m+1)$-tuple

$$(p;\widetilde{z}^1,\dots,\widetilde{z}^m;F^1,\dots,F^m)$$

with the described properties.

Definition 7.1. *Accordingly an **evolution of economies with temporarily fixed prices, quantity constraints and effective demand à la Benassy** is formally described by a $(2m+1)$-tuple of one-parametrizations*

$$(p_s;\widetilde{z}_s^1,\dots,\widetilde{z}_s^m;F_s^1,\dots,F_s^m)_{s\in[0,1]}$$

which are continuous in the natural sense such that any s-state forms an economy as defined above. In other words, p_s is a continuous path in $\mathbb{R}_+^n \setminus \{0^n\}$, and $\widetilde{z}_s^i$ and F_s^i are continuous one-parametrizations of functions in the usual sense.

An *equilibrium* of an economy is prevailing when for each agent the realized demands and supplies on all markets are consistent with the previously perceived rationing bounds on which he based his effective demand/supply signals. In other words, in an equilibrium no plan revisions are necessary since the supposed bounds for all individual plans turn out to be correct. To put it formally:

Definition 7.2. *For an economy* $(p; (\widetilde{z}^a)_{a=1,\ldots,m}; (F^a)_{a=1,\ldots,m})$ *a system of rationing bounds* $(\underline{z}^a, \overline{z}^a)_{a=1,\ldots,m}$ *with* $\underline{z}^a \in G_-, \overline{z}^a \in G_+$ *is an **equilibrium** if the following holds:*

$$\text{if } \widetilde{z}_i^a(p; \underline{z}^a, \overline{z}^a) < \underline{z}_i, \text{ then } F_i^a(\widetilde{z}^1, \ldots, \widetilde{z}^m) = \underline{z}_i^a,$$

$$\text{if } \widetilde{z}_i^a(p; \underline{z}^a, \overline{z}^a) > \overline{z}_i, \text{ then } F_i^a(\widetilde{z}^1, \ldots, \widetilde{z}^m) = \overline{z}_i^a,$$

and if $\underline{z}_i^a \leq \widetilde{z}_i^a(p; \underline{z}^a, \overline{z}^a) \leq \overline{z}_i^a$, *then* $F_i^a(\widetilde{z}^1, \ldots, \widetilde{z}^m) = \widetilde{z}_i^a(p; \underline{z}^a, \overline{z}^a)$ *for every* $i = 1, \ldots, n, \ a = 1, \ldots, m.$

(ii) As a reasonable candidate for an equilibrium equivalent self-mapping we propose the following intuitive mapping

$$g : (G_-)^m \times (G_+)^m \longrightarrow \mathbb{R}^{2nm}$$

$$\begin{pmatrix} \underline{z}^1 \\ \vdots \\ \underline{z}^m \\ \overline{z}^1 \\ \vdots \\ \overline{z}^m \end{pmatrix} = \begin{pmatrix} \begin{pmatrix} \underline{z}_1^1 \\ \vdots \\ \underline{z}_n^1 \end{pmatrix} \\ \vdots \\ \begin{pmatrix} \underline{z}_1^m \\ \vdots \\ \underline{z}_n^m \end{pmatrix} \\ -- \\ \begin{pmatrix} \overline{z}_1^1 \\ \vdots \\ \overline{z}_n^1 \end{pmatrix} \\ \vdots \\ \begin{pmatrix} \overline{z}_1^m \\ \vdots \\ \overline{z}_n^m \end{pmatrix} \end{pmatrix} \mapsto \begin{pmatrix} \begin{pmatrix} l_1^1 \\ \vdots \\ l_n^1 \end{pmatrix} \\ \vdots \\ \begin{pmatrix} l_1^m \\ \vdots \\ l_n^m \end{pmatrix} \\ -- \\ \begin{pmatrix} u_1^1 \\ \vdots \\ u_n^1 \end{pmatrix} \\ \vdots \\ \begin{pmatrix} u_1^m \\ \vdots \\ u_n^m \end{pmatrix} \end{pmatrix}$$

where

$$l_i^a = \begin{cases} \underline{z}_i^a + F_i^a(\widetilde{z}^1(p; \underline{z}^1, \overline{z}^1), \ldots, \widetilde{z}^m(p; \underline{z}^m, \overline{z}^m)) \\ \quad - \max(\underline{z}_i^a; \widetilde{z}_i^a(p; \underline{z}^a, \overline{z}^a)) & \text{for } \widetilde{z}_i^a(p; \underline{z}^a, \overline{z}^a) \le 0. \\ \underline{z}_i^a & \text{for } \widetilde{z}_i^a(p; \underline{z}^a, \overline{z}^a) \ge 0 \end{cases}$$

and

$$u_i^a = \begin{cases} \overline{z}_i^a & \text{for } \widetilde{z}_i^a(p; \underline{z}^a, \overline{z}^a) \le 0. \\ \overline{z}_i^a + F_i^a(\widetilde{z}^1(p; \underline{z}^1, \overline{z}^1), \ldots, \widetilde{z}^m(p; \underline{z}^m, \overline{z}^m)) \\ \quad - \min(\overline{z}_i^a; \widetilde{z}_i^a(p; \underline{z}^a, \overline{z}^a)) & \text{for } \widetilde{z}_i^a(p; \underline{z}^a, \overline{z}^a) \ge 0 \end{cases}$$

with $i = 1, \ldots n$, and $a = 1, \ldots, m$.

We have to verify several things.

(1) g is *continuous*. This follows directly from the assumption that realized transactions must be voluntary, i.e.

$$0 \le |F_i^a(p; \widetilde{z}^1, \ldots, \widetilde{z}^m)| \le |\widetilde{z}_i^a(p; \underline{z}^a, \overline{z}^a)|.$$

The following Figure 7.1 gives an illustration. The thick line is the graph

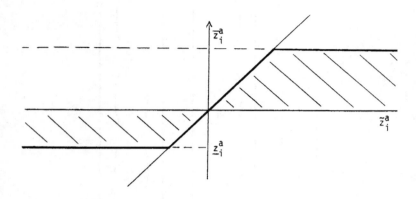

Fig. 7.1: Continuity of the Equilibrium Equivalent Self-Mapping

of the function

$$\widetilde{z}_i^a \mapsto \begin{cases} \max(\underline{z}_i^a; \widetilde{z}_i^a) & \text{for } \widetilde{z}_i^a \le 0 \\ \min(\overline{z}_i^a; \widetilde{z}_i^a) & \text{for } \widetilde{z}_i^a \ge 0. \end{cases}$$

Accordingly, the closure of the shaded areas shows the range of F_i^a.

(2) That the fixed-point set of g equals the set of equilibria of an economy is obvious from the definitions.

(3) It remains to verify that g actually is a self-mapping, i.e. is a mapping into $(G_-)^m \times (G_+)^m$.

To show this let us start with the l_i^a-components of g. We have to show that

$$-b_i \le \underline{z}_i^a + F_i^a(\widetilde{z}^1, \ldots, \widetilde{z}^m) - \max(\underline{z}_i^a; \widetilde{z}_i^a(\underline{z}^a, \overline{z}^a)) \le 0.$$

The *second* relation is an immediate consequence of $\underline{z}_i^a \le 0$ and $F_i^a(\ldots) \le 0$, and

$$0 \ge \max(\underline{z}_i^a; \widetilde{z}_i^a(\underline{z}^a, \overline{z}^a)).$$

To verify the *first* relation we have to distinguish the following two cases:

(I) Assume that $F_i^a(\ldots) \le \underline{z}_i^a$. From the assumptions then follows that

$$\widetilde{z}_i(\ldots) \le F_i^a(\ldots) \le \underline{z}_i^a \le 0$$

$$\text{and } F_i^a(\ldots) \ge -b_i.$$

Then $\max(\underline{z}_i^a; \widetilde{z}_i^a(\ldots)) = \underline{z}_i^a$. Consequently,

$$\underline{z}_i^a + F_i^a(\ldots) - \max(\underline{z}_i^a; \widetilde{z}_i^a(\ldots)) = \underline{z}_i^a + F_i^a(\ldots) - \underline{z}_i^a = F_i^a(\ldots) \ge -b_i.$$

(II) Assume that $F_i^a(\ldots) > \underline{z}_i^a$. Then we have

$$\widetilde{z}_i^a(\ldots) \leq F_i^a(\ldots) \leq 0, \text{ and } -b_i \leq \underline{z}_i^a.$$

We have to make a further distinction of subcases:

(IIa) $\max(\underline{z}_i^a; \widetilde{z}_i^a(\ldots)) = \widetilde{z}_i^a(\ldots)$. Then $\underline{z}_i^a + [F_i^a(\ldots) - \widetilde{z}_i^a] \geq \underline{z}_i^a \geq -b_i$ since the term in the square brackets is non-negative, and we are done.

(IIb) $\max(\underline{z}_i^a; \widetilde{z}_i^a(\ldots)) = \underline{z}_i^a$. Then $\underline{z}_i^a + F_i^a(\ldots) - \underline{z}_i^a = F_i^a(\ldots) \geq -b_i$, and we are done.

□

The following Figure 7.2 depicts the situation on the real line:

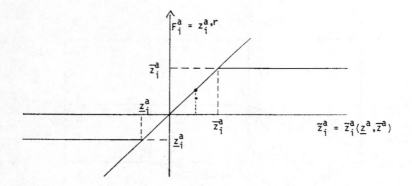

Fig. 7.2: Equilibrium Equivalent Self-Mapping Characteristics

Obviously, the corresponding problem with the u_i^a-components of our mapping g is completely symmetric: just reflect the whole problem from the negative real half line onto the positive one (cf. Figures 7.1 and 7.2).

(iii) From the definition of g it is immediately clear that an evolution of economies in the presented basic set-up in fact induces a continuous one-parametrization of associated equilibrium equivalent self-mappings as desired.

7.2 Evolutions Based on a Quantity Constrained Multi-Sectoral Model

(i) Our model economy to presented now is made up by m *sectors* (industries). Agents are households or firms. Each sector i, $i = 1, \ldots, m$, is completely characterized by its *sectoral labor market*, its *sectoral commodity market* and the interplay of the two markets. The activities of the agents in each sector i are as follows: sector-specific labor l_i is demanded by the firms of the sector and is supplied by the households of the economy. Households and firms demand the commodity of sector i, which is supplied by the firms of sector i. The prices for the m commodities $y_1, \ldots, y_m$ and for the m sector-specific labor types $l_1, \ldots, l_m$ are temporarily fixed at levels $(p_1, \ldots, p_{2m}) =: p$. In each sector i the temporarily fixed price-wage-system determines the *aggregate notional (Walrasian) demand intentions for labor and for the commodity*, $l_i^{d^*}, y_i^{d^*}$, and the *aggregate notional supply intentions* $l_i^{s^*}, y_i^{s^*}$. These notional plans are represented in the box diagram of Figure 7.3 by the two points $F_i = (l_i^{d^*}, y_i^{s^*})$ and $H_i = (l_i^{s^*}, y_i^{d^*})$. To simplify notation the vector p will be subsequently suppressed.

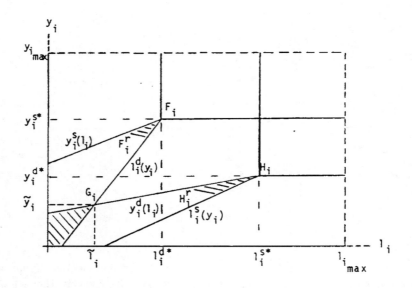

Fig. 7.3: Notional Quantity Constrained Supply and Demand Intentions in Sector i

For our later analysis we need the further assumption that all notional supply and demand intentions are bounded from above. This means that there are $2m$ positive real numbers $l_{1\max}, \ldots, l_{m\max}; y_{1\max}, \ldots, y_{m\max}$ which give the upper bounds of physical availability of the corresponding commodity or labor type. Evidently, this is not a restrictive assumption. Note that F_i, as usual, represents the *notional plans of the firms of sector* i, whereas H_i now represents the notional labor supply intentions of the households on sector i and the notional commodity demand intentions of the households on sector i and firms of the economy under the given temporarily fixed price-wage system.

The *inner-sectoral spillovers* are modelled in complete analogy to the well-known one-sectoral macromodel. The commodity supply reactions of the firms of sector i to perceived quantity constraints l_i on the sectoral labor market are represented by a reaction function $y_i^s(l_i)$. Correspondingly, the labor demand reactions of the sectoral firms to quantity constraints on the sectoral commodity market are represented by a function $l_i^d(y_i)$. Together the functions $y_i^s(l_i)$ and $l_i^d(y_i)$ form the *"reaction wedge"* F_i^r. Accordingly, the aggregated commodity demand reactions $y_i^d(l_i)$ of households to labor supply constraints on sector i, and the labor supply reactions $l_i^s(y_i)$ to commodity demand constraints together form the *"reaction wedge"* H_i^r. The reaction functions $y_i^s(-)$ and $y_i^d(-)$ are assumed to be continuous, weakly monotonically increasing functions of sectoral labor constraints l_i, and the reaction functions $l_i^s(-)$ and $l_i^d(-)$ are continuous, weakly monotonically increasing functions of sectoral commodity constraints y_i such that the intersection of the two wedges H_i^r and F_i^r forms a non-empty wedge with a unique vertex. This corresponds to the well-behavedness of the quantity constrained macromodel usually adopted in the literature (cf. Böhm 1982, 1989, for example).

In other words, in each sector firms plan to reduce their output if they become rationed on the sectoral labor market, and plan to reduce their labor input if they cannot realize their planned commodity sales. The possibility to hold inventories leads to the wedge. On the other hand, in each sector households will reduce (maintain) their labor supply by substituting leisure for labor if the commodity which is produced in the sector is (is not) a consumer good and demand is rationed on the commodity market. If it is (is not) a consumer good, households will reduce (maintain) their demand for the commodity of the sector if they get rationed on the sectoral labor market. The wedge is due to the possibility of saving.

Particularly, all supply intentions $y_i^s(l_i)$ and $l_i^s(y_i)$ must be technologically and physically feasible for the economy. (For the derivation from optimization programs cf. the literature on the subject, particularly Böhm 1982, 1989.)

The unique *sectoral quantity constrained temporary equilibrium* for sector i is given by the well determined vertex $G_i = (\widetilde{l}_i; \widetilde{y}_i)$ of the (entirely shaded) *intersection wedge* of the two reaction wedges F_i^r and H_i^r. (Note that the example drawn in Figure 7.3 shows the case which is analogous to the *Keynesian regime* of the well known macromodel).

At this point one might object the following: "while the firms' reaction functions $l_i^d(y_i)$ and $y_i^s(l_i)$ express well-founded causal relationships, the reaction functions of the H_i^r-wedge do not." Actually, it is not hard to think of examples where the overall demand for the product of a sector is not sensitively related to employment in this very sector, and where on the other hand the intentions of households to supply type-i-labor are not sensitive to demand rationing on the commodity market of this sector. But the reader should note that this is not inconsistent with our model, since *constant reaction functions* are admitted. Moreover, *all inter-sectoral constraint spillovers* will be incorporated in the next step. (Furthermore, *constant* reaction functions strongly favour equilibrium adjustment processes which are naturally derived from the well-known quantity tâtonnement process of the macromodel. The reader interested in details is referred to Lehmann-Waffenschmidt, 1987, Section 4.)

The following Figure 7.4 schematically summarizes the four well known *macro-rationing regimes* (phases) together with the five intermediate cases. The usual regime terminology shall also be maintained – here now characterizing, however, the state in each sector.

Clearly, our depicted regime figures are prototypes in so far as $l_i^s(y_i)$ and $l_i^d(y_i)$, or $y_i^s(l_i)$ and $y_i^d(l_i)$ respectively, may have points in common without violating our assumptions.

Thus, for our model with m sectors there are 9^m different combinations of sectoral rationing regimes possible.

Unfortunately, so far the model obviously is still unsatisfactory since each sector is isolated from any possible quantity rationing signals from the *other sectors*. In fact, up to now the model has the flavour of a partial analysis. Let us now show how to remedy this deficiency.

Our proposal to take the *inter-sectoral spillovers* into account is to continuously parametrizing the 4 reaction functions $l_i^s(y_i), l_i^d(y_i), y_i^s(l_i), y_i^d(l_i)$ of each sector i with constraint signals $\bar{l}_j$ and $\bar{y}_j$ from $[0, l_{j_{\max}}]$ and $[0, y_{j_{\max}}]$ transmitted from the other sectors. Thus, one obtains $4m$ continuous functions

$$l_i^s(\bar{l}_1, \ldots, \hat{l}_i, \ldots, \bar{l}_m; \bar{y}_1, \ldots, y_i, \ldots, \bar{y}_m)$$
$$l_i^d(\bar{l}_1, \ldots, \hat{l}_i, \ldots, \bar{l}_m; \bar{y}_1, \ldots, y_i, \ldots, \bar{y}_m) \tag{7.1}$$
$$y_i^s(\bar{l}_1, \ldots, l_i, \ldots, \bar{l}_m; \bar{y}_1, \ldots, \hat{y}_i, \ldots, \bar{y}_m)$$
$$y_i^d(\bar{l}_1, \ldots, l_i, \ldots, \bar{l}_m; \bar{y}_1, \ldots, \hat{y}_i, \ldots, \bar{y}_m), \quad i = 1, \ldots, m,$$

where l_i^s and l_i^d are continuous functions from the sub-cuboid $[0, l_{1_{\max}}] \times \ldots \times [0, l_{i-1_{\max}}] \times [0, l_{i+1_{\max}}] \times [0, l_{m_{\max}}] \times [0, y_{1_{\max}}] \times \ldots \times [0, y_{m_{\max}}]$ into $[0, l_{i_{\max}}]$ (denote $C^{2m} := [0, l_{1_{\max}}] \times \ldots \times [0, l_{m_{\max}}] \times [0, y_{1_{\max}}] \times \ldots \times [0, y_{m_{\max}}]$). y_i^s and y_i^d are continuous functions from $[0, l_{1_{\max}}] \times \ldots \times [0, l_{m_{\max}}] \times [0, y_{1_{\max}}] \times \ldots \times [0, y_{i-1_{\max}}] \times [0, y_{i+1_{\max}}] \times \ldots \times [0, y_{m_{\max}}]$ into $[0, y_{i_{\max}}]$. We will assume that for each i and any parametrizing constraint tuple $(\bar{l}_1, \ldots, \hat{l}_i, \ldots, \bar{l}_m; \bar{y}_1, \ldots, \hat{y}_i, \ldots, \bar{y}_m)$ the functions (7.1) form one of the rationing regimes compiled by Figure 7.4. Equivalently, we can formally represent the system (7.1) by m pairs of parametrized reaction wedges

$$F_i^r(\bar{l}_1, \ldots, \hat{l}_i, \ldots, \bar{l}_m; \bar{y}_1, \ldots, \hat{y}_i, \ldots, \bar{y}_m)$$
$$H_i^r(\bar{l}_1, \ldots, \hat{l}_i, \ldots, \bar{l}_m; \bar{y}_1, \ldots, \hat{y}_i, \ldots, \bar{y}_m), \quad i = 1, \ldots, m. \tag{7.2}$$

Up to now, the question is still open which signals are perceived by each sector as parametrizing constraint signals from the other sectors. Generally spoken these will be the (l, y)-coordinates of the states of the other sectors. However, whether these states are the sectoral equilibria or are any other states we will not decide in this study. We can do so since this question has no relevance for the equilibrium analysis we are purposed to do here.

Thus, under a given tuple of parametrizing quantity constraint signals from outside, i.e. *ceteris paribus*, the agents in each sector plan their reactions to inner-sectoral quantity constraints as in the traditional macromodel version. In particular, the geometrical representation in two dimensions familiar from the macromodel can be maintained (see below). More specifically, we now have m 2-dimensional parametrized diagrams characterizing to economy. Clearly, like in the unparametrized case above all demand and supply reactions of the fully parametrized model (7.1) must also be feasible.

As it is usual for the macromodel (see e.g. Böhm 1982), also in this study *uniqueness of sectoral quantity constrained equilibrium* is assumed for every sector and every parametrizing tuple. For our later purposes we furthermore need the following intuitive assumption (see remark (3) in Subsection (ii) below for a discussion): *for every sector i, $i \in \{1, \ldots, m\}$, the unique sectoral quantity constrained temporary equilibrium, i.e. the vertex*

$$
(\widetilde{l}_i(\overline{l}_1, \ldots, \hat{l}_i, \ldots, \overline{l}_m; \overline{y}_1, \ldots, \hat{y}_i, \ldots, \overline{y}_m);
$$
$$
\widetilde{y}_i(\overline{l}_1, \ldots, \hat{l}_i, \ldots, \overline{l}_m; \overline{y}_1, \ldots, \hat{y}_i, \ldots, \overline{y}_m))
$$

of the intersection wedge of the two parametrized reaction wedges

$$
F_i^r(\overline{l}_1, \ldots, \hat{l}_i, \ldots, \overline{l}_m; \overline{y}_1, \ldots, \hat{y}_i, \ldots, \overline{y}_m) \quad and
$$
$$
H_i^r(\overline{l}_1, \ldots, \hat{l}_i, \ldots, \overline{l}_m; \overline{y}_1, \ldots, \hat{y}_i, \ldots, \overline{y}_m),
$$

moves continuously in $[0, l_{i_{\max}}] \times [0, y_{i_{\max}}] \subset \mathbb{R}_+^2$ when the parametrizing $(2m-2)$-tuple $(\overline{l}_1, \ldots, \hat{l}_i, \ldots, \overline{l}_m; \overline{y}_1, \ldots, \hat{y}_i, \ldots, \overline{y}_m)$ is changed continuously in its domain cuboid

$$
[0, l_{1_{\max}}] \times \ldots \times [0, l_{i-1_{\max}}] \times [0, l_{i+1_{\max}}] \times \ldots \times [0, l_{m_{\max}}] \times
$$
$$
[0, y_{1_{\max}}] \times \ldots \times [0, y_{i-1_{\max}}] \times [0, y_{i+1_{\max}}] \times \ldots \times [0, y_{m_{\max}}] \subset \mathbb{R}_+^{2m-2}.
$$

To sum up, an *economy* in our present multi-sectoral quantity constrained set-up is specified by a system (7.1).

The total interdependence of the reaction functions of an economy (7.1) can *geometrically* be visualized by parametrizing the reaction functions in the i-th sector box-diagram by a constraint $(2m - 2)$-tuple $(\overline{l}_1, \ldots, \hat{l}_i, \ldots, \overline{l}_m; \overline{y}_1, \ldots, \hat{y}_i, \ldots, \overline{y}_m)$. Thus, in each of the m two-dimensional sectoral box-diagrams the reaction functions are shifted when the constraint signals from the other sectors change. To support intuition let us look at an example:

Households may reduce their demand $y_i^{d^*}(-)$ on the commodity market of the i-th sector if supply rationing on the labor market of some other sector j is strengthened. At the same time those households may increase their supply $l_i^{s^*}(-)$ on the labor market of the i-th sector who are also skilled for labor of type i. The two reaction functions $l_i^s(-; -, y_i, -)$ and $y_i^d(-, l_i, -; -)$ which start from $H_i(-) = (l_i^{s^*}(-); y_i^{d^*}(-))$ are changed accordingly. Geometrically this results in a right-downwards shifting of the H_i^r-wedge (cf. Figure 7.3).

On the other hand, the firms of sector i may reduce their supply $y_i^{s^*}(-)$, if the firms of sector j are labor-constrained and the commodity of sector j is needed for production in sector i. If there are firms in sector i which use technologies with fixed proportions of labor and commodity inputs, also labor demand $y_i^{d^*}(-)$ in sector i will diminish. This will result in a left-downwards shifting of the F_i^r-wedge.

Let us briefly summarize the economic characteristics of the model. It interlinks *internal planning* of the economic agents in each sector which only takes account of inner-sectoral constraint signals with *externally determined planning* depending on the signals from the other sectors. More precisely, there is a hierarchy of these two principles of planning: the sector-internal planning is completely carried out under every perceived tuple of sector-external signals. Thus, the model has a feature of "parametrized", or "damped", interdependency. Or to say it in other words, in each sector the agents plan as regards the constraint signals from the other sectors.

In general, in each sector i any shiftings of the two sectoral reaction wedges will also lead to a displacement of the sectoral quantity constrained temporary equilibrium G_i, i.e., of the vertex of the intersection wedge of the wedges F_i^r and H_i^r (cf. Figure 7.3). Formally, we write

$$
\begin{aligned}
G_i = {} & G_i[F_i^r(\bar{l}_1,\ldots,\hat{l}_i,\ldots,\bar{l}_m;\overline{y}_1,\ldots,\hat{y}_i,\ldots,\overline{y}_m), \\
& H_i^r(\bar{l}_1,\ldots,\hat{l}_i,\ldots,\bar{l}_m;\overline{y}_1,\ldots,\hat{y}_i,\ldots,\overline{y}_m)] \\
= {} & G_i(\bar{l}_1,\ldots,\hat{l}_i,\ldots,\bar{l}_m;\overline{y}_1,\ldots,\hat{y}_i,\ldots,\overline{y}_m) \\
= {} & (\widetilde{l}_i(\bar{l}_1,\ldots,\hat{l}_i,\ldots,\bar{l}_m;\overline{y}_1,\ldots,\hat{y}_i,\ldots,\overline{y}_m); \\
& (\widetilde{y}_i(\bar{l}_1,\ldots,\hat{l}_i,\ldots,\bar{l}_m;\overline{y}_1,\ldots,\hat{y}_i,\ldots,\overline{y}_m)).
\end{aligned}
\tag{7.3}
$$

Let us now come to the *notion of an equilibrium* of an economy (7.1) in the present set-up. Obviously, there is only one reasonable way to define a *multi-constrained general equilibrium*, or say an *equilibrium* for short, namely by a $2m$-tuple

$$
\begin{aligned}
& (\widetilde{l}_1,\ldots,\widetilde{l}_m;\widetilde{y}_1,\ldots,\widetilde{y}_m) = \\
& (\widetilde{l}_1(\widetilde{l}_2,\ldots,\widetilde{l}_m;\widetilde{y}_2,\ldots,\widetilde{y}_m),\ldots,\widetilde{l}_m(\widetilde{l}_1,\ldots,\widetilde{l}_{m-1};\widetilde{y}_1,\ldots,\widetilde{y}_{m-1}); \\
& \widetilde{y}_1(\widetilde{l}_2,\ldots,\widetilde{l}_m;\widetilde{y}_2,\ldots,\widetilde{y}_m),\ldots,\widetilde{y}_m(\widetilde{l}_1,\ldots,\widetilde{l}_{m-1};\widetilde{y}_1,\ldots,\widetilde{y}_{m-1})) \in \\
& C^{2m} \subset \mathbb{R}_+^{2m}
\end{aligned}
$$

where each pair

$$
\begin{aligned}
(\widetilde{l}_i;\widetilde{y}_i) = {} & (\widetilde{l}_i(\bar{l}_1,\ldots,\hat{l}_i,\ldots,\bar{l}_m;\widetilde{y}_1,\ldots,\hat{y}_i,\ldots,\widetilde{y}_m); \\
& \widetilde{y}_i(\bar{l}_1,\ldots,\hat{l}_i,\ldots,\bar{l}_m;\widetilde{y}_1,\ldots,\hat{y}_i,\ldots\widetilde{y}_m)) \in \mathbb{R}_+^2
\end{aligned}
$$

for $i = 1,\ldots,m$ denotes the unique sectoral quantity constrained temporary equilibrium of sector i. Thus, in an equilibrium the m sectoral quantity constrained equilibria $(\tilde{l}_i; \tilde{y}_i)$ are simultaneously mutually consistent. Or, in other words, all sectoral equilibrium signals are just mutually reproduced by the reactions of the agents when the reaction functions in each sector become *reparametrized* by the sectoral equilibrium values of the other sectors.

Definition 7.3. *An* **evolution of economies in the presented multi-sectoral quantity constrained basic set-up** *is given in the natural way by a 4m-tuple of continuous one-parametrizations of sectoral reaction functions*

$$(l_{1_s}^s(-), l_{1_s}^d(-), y_{1_s}^s(-), y_{2_s}^d(-); \ldots; l_{m_s}^s(-), l_{m_s}^d(-), y_{m_s}^s(-), y_{m_s}^d(-))_{s \in [0,1]}$$

such that any s-state forms an economy.

It is noteworthy that in contrast to the other basic models here any evolution can be geometrically visualized.

(ii) Constructing an equilibrium equivalent self-mapping for a given economy is straightforward.

Choose an arbitrary $i \in \{1, \ldots, m\}$ and an arbitrary admissible parametrizing constraint tuple $(\bar{l}_1, \ldots, \hat{l}_i, \ldots, \bar{l}_m; \bar{y}_1, \ldots, \hat{y}_i, \ldots, \bar{y}_m)$ for sector i. Dropping the parametrizing constraint tuple for the moment in order to simplify notation the sectoral reaction functions $y_i^s(l_i), l_i^d(y_i), l_i^s(y_i)$, and $y_i^d(l_i)$ can be extended beyond the vertices F_i and H_i by constant functions up to the upper bounds $l_{i_{\max}}$ and $y_{i_{\max}}$ respectively. This is indicated in the box-diagram of Figure 7.3 by broken segments.

Re-establishing now the parametrization notation we propose the following natural mapping from the $2m$-dimensional compact cuboid

$$C^{2m} = [0, l_{1_{\max}}] \times \ldots \times [0, l_{m_{\max}}] \times [0, y_{1_{\max}}] \times \ldots \times [0, y_{m_{\max}}] \subset \mathbb{R}_+^{2m}$$

into itself:

$$G : (\bar{l}_1, \ldots, \bar{l}_m; \bar{y}_1, \ldots, \bar{y}_m) \mapsto [\tilde{l}_1(\bar{l}_2, \ldots, \bar{l}_m; \bar{y}_2, \ldots, \bar{y}_m), \ldots,$$
$$\tilde{l}_m(\bar{l}_1, \ldots, \bar{l}_{m-1}; \bar{y}_1, \ldots, \bar{y}_{m-1});$$
$$\tilde{y}_1(\bar{l}_2, \ldots, \bar{l}_m; \bar{y}_2, \ldots, \bar{y}_m), \ldots,$$
$$\tilde{y}_m(\bar{l}_1, \ldots, \bar{l}_{m-1}; \bar{y}_1, \ldots, \bar{y}_{m-1})].$$

Each pair $(\tilde{l}_i(\bar{l}_1, \ldots, \hat{l}_i, \ldots, \bar{l}_m; \bar{y}_1, \ldots, \hat{y}_i, \ldots, \bar{y}_m);$ $\tilde{y}_i(\bar{l}_1, \ldots, \hat{l}_i, \ldots, \bar{l}_m; \bar{y}_1, \ldots, \hat{y}_i, \ldots, \bar{y}_m))$ denotes the *sectoral constrained equilibrium of sector i* as defined in (4.2.2), i.e. the vertex of the intersection wedge of the two sectoral reaction wedges $H_i^r(-)$ and $F_i^r(-)$, under the parametrizing $(2m-2)$-tuple $(\bar{l}_1, \ldots, \hat{l}_i, \ldots, \bar{l}_m; \bar{y}_1, \ldots, \hat{y}_i, \ldots, \bar{y}_m))$.

Clearly, the properties of the mapping G are essentially determined by the properties of the economy. It follows immediately from the definition and the assumptions that G is a well-defined continuous self-mapping of C^{2m} whose fixed-point set equals the set of equilibrium of the economy.

Finally, some remarks are in order.

(1) There is some noteworthy peculiarity with this model concerning the equality of the fixed-point set and the equilibrium set which at the first glance seems to produce a paradox: assume that we are given a certain parametrization of the m sector diagrams by a parameter tuple $(l_1, \ldots, l_m; y_1, \ldots, y_m)$ so that the corresponding m sectoral equilibria constitute a *fixed point of the mapping* G. Actually, this generally does not mean that re-parametrizing the reaction functions in each sector diagram by the coordinates of the sectoral equilibria of the other sectors necessarily leaves the reaction functions unchanged. What it means is just that no changes of any reaction functions are possible which also change the position of any sectoral equilibrium. However, a further re-re-parametrization does not change the reaction functions anymore. These considerations also apply to the converse case when a configuration of the reaction functions in the m sector diagrams is given whose sectoral equilibria form a *multi-constrained general equilibrium*.

(2) It might be argued that for each sector i the coordinates of the sectoral equilibria which are perceived as parametrizing constraints by the other sectors may well be a priori restricted to a certain subset $C_i \subset [l_i', l_i''] \times [y_i', y_i'']$ with $0 \leq l_i' \leq l_i'' \leq l_{i_{\max}}$ and $0 \leq y_i' \leq y_i'' \leq y_{i_{\max}}$. Accordingly, the system (7.2) and the mapping G should be restricted to $\widetilde{C} := C_1 \times \ldots \times C_m \subset C^{2m}$. Clearly, $G|_{\widetilde{C}}$ is a self-mapping of $\widetilde{C}$. The reader may easily note that our construction of an equilibrium equivalent self-mapping also works for this case if $G|_{\widetilde{C}} : \widetilde{C} \longrightarrow \widetilde{C}$ can be continuously extended to a mapping $\overline{G} : C^{2m} \longrightarrow \widetilde{C}$. Thus, all fixed points are in $\widetilde{C}$. A *simple example* for this is given by C_i being an interval for each $i = 1, \ldots, m$. In this case Brouwer's Theorem can directly be applied to $G|_{\widetilde{C}}$.

(3) Our main assumption only requires that in each sector the *sectoral equilibrium moves continuously* with continuously changing parametrizing constraint signals form the other sectors. Particularly, it does not require that also the vertices of the reaction wedges move continuously. If one thinks of the parametrized reaction functions in each sector as derived from four *aggregate* sectoral maximization programs, continuous movements of the parametrized reaction wedges in each sector i mean that the four objective functions must be sufficiently well behaved so that continuous parameter changes in fact lead to continuous changes of maximizing arguments $l_i^{s^*}(-)$, $l_i^{d^*}(-)$, $y_i^{s^*}(-)$, $y_i^{d^*}(-)$ and $l_i^s(-;-,y_i,-)$, $l_i^d(-;-,y_i,-)$, $y_i^s(-,l_i,-;-)$, and $y_i^d(-,l_i,-;-)$. (For instance this is the case if shortages in a sector are continuously 'redistributed' to the other sectors.) On the other hand, if one thinks of the sectoral reaction functions as being the outcome of *individual* decisions, their continuous movements can be justified by aggregation. On the whole, continuous movements of the sectoral reaction functions do not appear to be more restrictive than the continuity of the reaction functions themselves in the traditional macromodel.

(4) Though by assumption in every sector there is always a *unique sectoral temporary equilibrium for any given parametrizing constraint tuple*, it may well happen that there are *multiple equilibria* for an economy. This means that also our multi-sectoral model has the familiar feature of indeterminacy as regards the equilibrium set. This can be illustrated by a graphical example:

Let us for instance choose $m = 5$, and accordingly ten real positive numbers as upper bounds $l_{1\max}, \ldots, l_{5\max}; y_{1\max}, \ldots, y_{5\max}$. Now draw for each sector a box-diagram (cf. Figure 7.3) so that the vertex $G_i = (\tilde{l}_i; \tilde{y}_i)$ of the intersection wedge of the reaction wedges F_i^r and H_i^r constitutes the unique sectoral equilibrium. Now choose an $i \in \{1, \ldots, 5\}$ and consider the i-th sector's diagram. Write the 8 $(= 2m - 2)$ parametrizing coordinates $(\tilde{l}_1, \ldots, \hat{l}_i, \ldots, \tilde{l}_5; \tilde{y}_1, \ldots, \hat{y}_i, \ldots, \tilde{y}_5)$ of the sectoral equilibria of the remaining 4 sectors at the i-th sector's reaction functions and at its points F_i, H_i, and $G_i = (\tilde{l}_i; \tilde{y}_i)$. Now draw a second series of 5 sectoral box-diagrams with different reaction functions and different sectoral equilibria $G_i' = (\tilde{l}_i'; \tilde{y}_i')$ and parametrize them in the way just described. Every system of $4 \cdot 5$ reaction functions

$$l_i^s(l_1, \ldots, \hat{l}_i, \ldots, l_5; y_1, \ldots, y_5)$$
$$l_i^d(l_1, \ldots, \hat{l}_i, \ldots, l_5; y_1, \ldots, y_5)$$
$$y_i^s(l_1, \ldots, l_5; y_1, \ldots, \hat{y}_i, \ldots, y_5)$$
$$y_i^d(l_1, \ldots, l_5; y_1, \ldots, \hat{y}_i, \ldots, y_5), \quad i = 1, \ldots, 5,$$

which admits these two situations has at least these two multi-constrained general equilibria. (Extension of this type of example to produce continua of equilibria is obvious.) Actually it is not hard to think of inner- and intersectoral interdependencies (i.e. shifting rules for the reaction functions) which are economically reasonable and lead to these general equilibria.

(iii) From the construction of the equilibrium equivalent self-mapping G and from the definition of an evolution of economies follows directly that an evolution of economies

$$(l_{1_s}^s(-), l_{1_s}^d(-), y_{1_s}^s(-), y_{1_s}^d(-); \ldots; l_{m_s}^s(-), l_{m_s}^d(-), y_{m_s}^s(-), y_{m_s}^d(-))_{s \in [0,1]}$$

induces a continuous one-parametrization of equilibrium equivalent self-mappings $(G_s)_{s \in [0,1]}$.

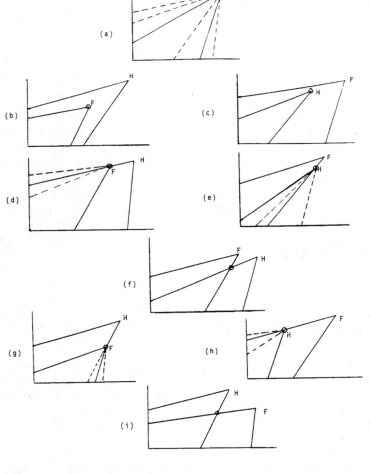

Fig. 7.4: The 9 Regime Types of the Quantity Constrained Multi-Sectoral Model

(a) Walrasian Case

(b) Regime of Classical Unemployment   (c) Regime of Under Consumption

(d) Intermediate Case (b)-(f)   (e) Intermediate Case (c)-(f)

(f) Keynesian Regime

(g) Intermediate Case (b)-(i)   (h) Intermediate Case (c)-(i)

(i) Regime of Repressed Inflation

(quantity constrained equilibria are marked by ∘, the broken lines indicate possible
alternative positions of reaction functions)

8

Conclusions

In Part I of this monograph the fundamentals for the subsequent formal and economic analysis have been laid. This was done by introducing and analyzing nine models of general equilibrium, which we have called 'basic models', or 'basic set–ups'. Some of them have been adopted from the literature (Chapters 4 and 6), while the others are new (Chapters 5 and 7). Thus the basic models have not been introduced as an end in themselves, but as basic set–ups for constructing the main subject of investigation, i.e., evolutions. All nine types of evolutions of economies have been constructed on the same formal principle as continuous one-parametrizations, or say one–parameter families, of economies. It has been emphasized that in the study *any successions* of states that satisfy our assumptions, not only evolutions of an economic system ongoing over historical time, have been admitted.

The findings in the subsection *iii* provide the basis for systematically and comprehensively tackling the fundamental question of how the equilibrium set changes when the underlying economy evolves continuously. In fact, this has become one of the major concerns of theoretical economics since the fundamental indeterminateness of the basic exchange framework was discovered in the early 1970s. More specifically we are searching for general structural regularities, or say 'laws', of the induced evolution of equilibria. Actually, the results in the following Part II will be fairly surprising to the reader who is familiar with the examples illustrating regular equilibrium theory, which strongly suggest that there is a maximal arbitrariness of the dependently evolving equilibrium set.

Part II

Formal Analysis

Introduction to Part II

Part II of the study provides the reader with our analytical results on evolutions of economies. All results are economically motivated and discussed. Nevertheless, their main economic significance will become clear through the applications in Part III of the study.

Before we start describing the approach and the results of Part II the concepts and findings of Part I are briefly recalled. Thus, the reader who is mainly interested in the analytical results of Part II can start here. Each chapter of Part I has been devoted to one model framework and has been organized according to the same scheme: *Each section* deals with one specific basic model and is further divided into *three subsections*. Subsection (i) motivates the model and specifies the notions of an economy, an equilibrium, and of an evolution of economies in this set–up. Subsection (ii) presents the construction of the most important technical tool of the analysis, a continuous self–mapping of some compact Euclidean space which equivalently transforms the zero–problem of the existence of equilibria into a fixed–point problem. This is done in order to make the equilibrium analysis amenable to the powerful tools of the *one–parametrized fixed–point theory* here in Part II. We will call the addressed mapping an 'equilibrium equivalent self–mapping'. Finally, subsection (iii) contains the verification that any admissible evolution of economies in the present basic set–up actually yields a *continuous* one-parametrization of associated equilibrium equivalent self–mappings. In fact, it is this property of an evolution of economies as we have formalized it that will turn out to be the essential prerequisite for the results in Part II. Fortunately, the technical work done in subsections (ii) and (iii) of the chapters in Part I has covered the major part of the technical effort necessary for the central analytical results in Chapters 10 and 11.

The central analytical result is given in Chapter 10. It ensures that the equilibrium set of any evolution considered in Part I has a certain structural property. Loosely speaking, this means that above any evolution of economies, there is a path of equilibria in the graph of the equilibrium correspondence which is either itself geometrically well-behaved, or which can be approximated by a well-behaved path. This result has also been given independently for the basic model of a large exchange economy by Mas-Colell (1985, Section 5.8). However, the methods used by Mas-Colell are completely different from ours. Comparing this result to the well-known indeterminateness result of the static exchange framework we see that the degree of indeterminateness on the one-parametrized level is significantly smaller than on the static level.

The existence of approximating "near"-equilibrium paths naturally raises a further question. How large is the subclass of those evolutions that already have well-behaved equilibrium paths in each basic set-up? Or, to formulate it more stringently, can any evolution be approximated by one that has a well-behaved equilibrium path?

In Chapter 11, it is shown that the affirmative answer to this question is right for each of the nine basic set-ups. For the exchange framework a closely related result has been shown by A. Mas-Colell using the concept of open-density of 'regular' evolutions. Nevertheless, here we provide constructive methods to achieve nice approximating evolutions, whereas Mas-Colell just gives an abstract existence result. From our constructions, we derive a further new result that shows that the equilibrium correspondence of each of the nine basic models introduced in Part I is extremely regularly connected. This result considerably extends the result of the manifold property of the graph of the Walras correspondence by Y. Balasko (see Balasko's various articles in the reference list).

Our conceptualization of an evolution of economies admits two obvious economic interpretations, which will be analyzed in Chapter 12. On the one hand, one may stress the aspect of *course*. Then an evolution describes an economy, which from its initial state economy, evolves somewhere in the space of economies – the only restriction being that it has to obey the weak conditions indicated in Part I. On the other hand, one may stress the aspect that an evolution *connects* its initial state economy to its terminal state economy. This interpretation immediately raises the question of whether there is always an evolution connecting any two given economies. In Section 12.1, we show that this is in fact possible.

When considering evolutions as evolutions in historical time, it seems to be most natural to also include new commodities entering the economy during its evolution and old commodities leaving it. The formal extensions necessary to achieve this in all basic set-ups are provided in Section 12.2.

However, the natural question remains whether there are possibly other structural properties of the equilibrium set of evolutions that also generally hold. In Chapter 13 a definitive answer will be given to this question for the basic models from the exchange framework (Chapters 4 and 5): there are no other general structural properties. Our result is the one-parametrized generalization of Mas-Colell's famous result on the non-restrictedness of the equilibrium set of a static exchange economy from 1977. Furthermore our result is to be seen as complementary to the related results on the local surjectiveness of the graph of the Walras correspondence by B. Allen (1981).

The gains and losses of our approach and our analytical results compared with related ones from the literature are discussed in detail in Chapter 14.

In a nutshell our results significantly extend those from regular equilibrium theory and those on the global properties of the graph of the Walras correspondence – both developed in the 1970ies and 1980ies – as far as these are related to ours.

10

Near-Equilibrium Paths

In this Chapter we will reap the first fruits of our technical efforts in Part I. Actually, our constructions in subsections (ii) and (iii) of the Chapters 4 to 7 put us in a position to show for any presented type of evolution that there is a certain general structure property of its equilibrium set.

To be more specific, in Section 10.1 we will prove that to each type of a continuous one-parametrization of equilibrium equivalent self-mappings constructed in Part I a certain result from algebraic topology applies which ensures the following: there is a *connected component of the equilibrium set of* the underlying evolution of economies which joins bottom and top of the homotopy space. (Loosely speaking we will also say that the connected component 'expands over the whole evolution'.) The existence of such a joining connected equilibrium component in turn ensures the existence of a nicely behaved joining path which approximates the equilibrium set of the evolution arbitrarily closely. This we will accordingly call a '*near-equilibrium path*'. Actually, it is the existence of near-equilibrium paths which proves to be crucial for our further applications in Part III of the monograph.

In principle the addressed result from algebraic topology has already been known since 1960 when Felix Browder published his paper "On continuity of Fixed Points under Deformation of Continuous Mappings". Browder, however, presented his result in a fairly general setting unfortunately obscuring his achievements somewhat. Applied to our situation his result, however, does not allow for boundary equilibria. Fortunately, there is an extended result by Dieter Puppe (1979, Corollary 5.6) which proves to be a more appropriate mathematical tool for our purposes.

Having achieved the general structure property of the existence of near-equilibrium paths for evolutions it is most natural to ask whether possibly there are still further general structure properties of the equilibrium sets of evolutions. We will settle this question in Chapter 13 for the exchange framework. Actually, we will be able to demonstrate that the existence of a joining connected equilibrium component is the only structure property which is generally valid.

The existence of at least one near-equilibrium path for any admissible evolution from Part I is ensured by Theorem 10.2 in Section 10.1. Theorem 10.10 in Section 10.2 gives an answer to the important question of how one can distinguish whether an arbitrary point in the homotopy space lies on a joining connected equilibrium component of a given evolution, or not. Moreover, the method which we will provide is particularly powerful for detecting *initial points* of joining connected equilibrium components.

10.1 Existence of Joining Equilibrium Components and of Near-Equilibrium Paths for Each Type of Evolution From Part I

We begin this Section with the precise definition of a near-equilibrium path.

Definition 10.1. *Let any admissible evolution $(\zeta_s)_{s\in[0,1]}$ in any basic model from Part I be given. Let be $X \in \{\Delta^{n-1}\backslash L, \mathring{\Delta}^{n-1}, \mathbb{R}_+^n, \mathbb{R}_+^n\backslash(\bigcup_{i=1}^n D_i), \Delta^{n-1} \times [0,\beta], \Delta^{n-1} \times [0,\beta] \times [0,1], (G_-)^m \times (G_+)^m, C^{2m}\}$ the corresponding domain. Then for any $\epsilon > 0$ an ϵ-near-equilibrium path for the evolution $(\zeta_s)_{s\in[0,1]}$ is a finitely piecewise linear path, or say a finitely polygonal path,*

$$\pi : [0,1] \longrightarrow X \times [0,1]$$

whose arc $\pi[0,1]$ lies in the ϵ-neighborhood of the equilibrium set of $(\zeta_s)_{s\in[0,1]}$ in $X \times [0,1]$ and furthermore joins the bottom $X \times \{0\}$ and the top $X \times \{1\}$ of $X \times [0,1]$.

The reader should be well aware that an ϵ-near equilibrium price path ϵ-approximates the equilibrium set of an evolution *on the whole*. In particular, this means that it *need not* ϵ-approximate every s-state equilibrium set when the equilibrium set of the whole evolution decomposes into several path components. Figure 10.1 below shows an example for this. Nevertheless, we will

Fig. 10.1: Approximation Pathology of an ε-Near Equilibrium Path

see that for a large class of evolutions any ϵ-near equilibrium path indeed also ϵ-approximates any s-state equilibrium set (Corollary 10.5).

Now we are ready to state the following result which will be central for our study.

Theorem 10.2. *For any admissible evolution $(\zeta_s)_{s\in[0,1]}$ basing on any one of the nine basic models from Part I there is at least one ϵ-near equilibrium path for any given $\epsilon > 0$.*

The following proof of Theorem 10.2 will not only provide the reader with the logical chain of mathematical arguments establishing the statement of the Theorem, but also with a detailed discussion on its meaning and the surroundings. This is also the reason why we will not relegate it to the appendix.

Proof. Fortunately, the major part of work has already been done in Part I. Actually, from the subsections (iii) of the Sections 4.1 to 7.2 follows that the given evolution $(\zeta_s)_{s\in[0,1]}$ induces a *continuous* one-parametrization of equilibrium equivalent self-mappings

$$(g_s)_{s\in[0,1]} : K \times [0,1] \longrightarrow K$$

with appropriate $K \in \{\overline{\Delta}^{n-1}, \overline{T}^n, \overline{\Delta}^{n-1} \times [0,\beta], \overline{\Delta}^{n-1} \times [0,\beta] \times [0,1], (G_-)^m \times (G_+)^m, C^{2m}\}$.

Let us now look at the properties of K. K is compact, and, particularly, it is a Euclidean neighborhood retract. Since K is furthermore contractible, it is also acyclic. Hence, any self-mapping of K has Lefschetz number $+1$ (see Brown (1971), II.c).

This means that we have posed ourselves in a situation to which the following result from one-parametrized algebraic topological fixed point theory applies:

Proposition 10.3. *Let K be a compact subset of $\mathbb{R}^n$ and a neighbourhood retract. Let $(g_s)_{s\in[0,1]} : K \times [0,1] \longrightarrow K$ be a continuous family of maps, and let F be the union of the fixed-points of the mappings g_s, i.e.,*

$$F := \bigcup_{s\in[0,1]} Fix\,(g_s) \subset K \times [0,1].$$

Then the fixed-point index λ of g_s equals the Lefschetz number of g_s, and is independent of s. If $\lambda \neq 0$, then F has a connected component C which meets bottom $K \times \{0\}$ and top $K \times \{1\}$ of the homotopy space.

We also need the following

Definition 10.4. *We will call a connected component of the equilibrium set of an evolution of economies which meets bottom and top of the homotopy space a **joining equilibrium component of the evolution**.*

The Proposition has been proven by Puppe (1979, Corollary 5.6). Apparently, the existence of a joining equilibrium component C for the evolution $(\zeta_s)_{s\in[0,1]}$ brings us more closely to our goal. However, such a connected joining equilibrium component may still display some bad geometrical features. Let us come back to this after the proof will be finished.

Before continuing the proof, however, a further remark on Proposition 10.3 is in order. A closely related result has been proven by F. Browder (1960, Theorem 2, p. 186). However, Browder uses a more restrictive boundary assumption. He does not admit fixed-points (x, s) on the boundary of $K \times [0, 1]$ (cf. also the corollary of Browder's result by Mas-Colell, 1974, Theorem 1). Actually, this result would not work for those of our basic models which admit boundary equilibria, i.e. the exchange model by Arrow/Hahn, the tax equilibrium models by Kehoe, and the models from the temporary quantity constrained equilibrium framework. Nevertheless, in another respect, which however is less important for our purposes, Browder's result is more general than Puppe's: K may be from a more general class of spaces.

Now, let us do the last step of our proof of Theorem 10.2 by demonstrating that there is a near-equilibrium path in any relatively open ϵ-neighborhood

$$\left[\bigcup_{x \in C} \overset{o\,n+1}{B_\epsilon} (x) \right] \cap (K \times [0, 1]) =: \bigcup_{x \in C} \overset{o\,n+1}{B_{\epsilon_r}} (x) =: C_\epsilon$$

of any joining component C of the equilibrium set of the given evolution $(\zeta_s)_{s \in [0,1]}$ (note that by definition a relatively open ϵ_r-ball is $\overset{o\,n+1}{B_{\epsilon_r}} (x) = \overset{o\,n+1}{B_\epsilon} (x) \cap (K \times [0, 1])$).

As C is compact, finitely many relatively open ϵ_r-balls $\overset{o\,n+1}{B_{\epsilon_r}} (x_1), \ldots, \overset{o\,n+1}{B_{\epsilon_r}} (x_k)$ of the ϵ-neighbourhood C_ϵ are sufficient to cover C. Denote their union by C_ϵ^f.

Now consider all pairs (x_i, x_j), $i \neq j$, of centers of the relative ϵ_r-balls $\overset{o\,n+1}{B_{\epsilon_r}} (x_i)$, and consider the graph g_C' consisting of all segments $\overline{x_i x_j}$ which *are contained in* C_ϵ^f. If one adds all segments to g_C' which are orthogonal to $\mathbb{R}_+^n \times \{0\}$ and connect a center $x_i \in \{x_1, \ldots, x_k\}$ with $\mathbb{R}_+^n \times \{0\}$ or with $\mathbb{R}_+^n \times \{1\}$ *within* $\overset{o\,n+1}{B_{\epsilon_r}} (x_i)$, one obtains a *finitely polygonal graph* g_C in C_ϵ^f which contains a near-equilibrium price path as desired. $\qquad \square$

In his monograph (1985, Proposition 5.8.2) Mas-Colell presents the analogue of our Theorem 10.2 for the basic model of an explicit finite exchange economy (for this notion cf. the Section 'Mathematical Preliminaries' above). Actually, Mas-Colell's method of deriving the result is quite different from ours.

We now proceed by pointing out the reasons why it is generally necessary to approximate a joining equilibrium component by near-equilibrium paths in order to get a *nicely behaved* path in the homotopy space. Let us use the notion of a 'nicely behaved path' for the moment in the intuitive geometric sense which means a particle moving along a nicely behaved path in a highly regular manner. Particularly, there should not occur any complicated movements like infinitely many oscillations for instance. Thus, a finitely piecewise linear near-equilibrium path is a *prototype* of a nicely behaved path. In Chapter 11 we will give an analytically precise and comprehensive characterization of the class of nicely behaved paths and of the class of evolutions which produce them.

Let us now look at the possible geometrically bad behavior of joining equilibrium components. First and foremost a joining equilibrium component *need not be path connected*. In other words, it may for instance contain parts like the closure of the graph of $\sin 1/x$. We will give an example of an evolution producing this below. However, even if a joining equilibrium component is path connected, it may well happen that some of its points can only be connected by paths with infinitely long arcs caused by infinitely many oscillations. The reader finds simple examples for this in the Mathematical Preliminaries at the beginning of our study.

Another example of a path whose arc is of finite length though it undergoes infinitely many oscillations is given by a "saw tooth path". It consists of infinitely many segments whose lengths can be estimated from above by the terms of a sequence which generates a convergent series.

Unfortunately, any of these complications actually *can occur* in the equilibrium set of an evolution. They even *cannot be removed by additional differentiability conditions* on the evolution. The following example makes this clear. ζ_0 is a smooth function with a linear part over $[y, z]$ (use the function

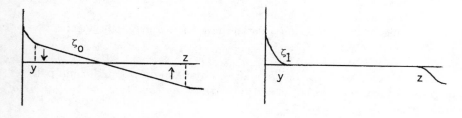

(a) Regular Equilibrium   (b) Continuum of Critical Equilibria

Fig. 10.2: Initial State and Final State of a Smoothly Oscillation Movement

$$x \mapsto \begin{cases} 0, & x \leq 0 \\ e^{-1/x^2}, & x > 0 \end{cases}$$ at the bends $\zeta_0(y)$ and $\zeta_0(z)$). Actually, the following

movement of ζ_0 yields a *smooth* one-parametrization $(\zeta_s)_{s \in [0,1]} : \mathbb{R}_+ \times [0, 1] \longrightarrow \mathbb{R}$; ζ_s is linear over $[y, z]$ for any s, and $\zeta_s(y)$ performs a damped oscillation whose time path looks like $x \cdot \sin \frac{1}{x}$. If $\zeta_s(z)$ correspondingly oscillates in counter-rhythm, this results in a *smoothly oscillating movement* $(\zeta_s)_{s \in [0,1]}$ with final state ζ_1 as in Figure 10.2. Thus, the trace $(G_s)_{s \in [0,1]}$ of the oscillating unique zero in the homotopy space $R_+ \times [0, 1]$ *looks like the closure of the graph of* $\sin \frac{1}{x}$.

From the construction of the last part of the proof of Theorem 10.1 follows immediately

Corollary 10.5. *If at least one joining equilibrium component of an evolution* $(\zeta_s)_{s\in[0,1]}$ *is even path connected, then for any* $\epsilon > 0$ *there is an* ϵ-*near equilibrium price path for* $(\zeta_s)_{s\in[0,1]}$ *which particularly also* ϵ-*approximates every s-state equilibrium set.*

Proof.   Just *exclude* from the construction of the finitely polygonal graph g_C in the final part of the proof of Theorem 5.1 all segments $\overline{x_i x_j}$ with the following property: the endpoints x_i and x_j cannot be connected by a path which lies *in* C *and in* $\overset{\circ}{B}{}^{n+1}_{\epsilon_r}(x_i) \bigcup \overset{\circ}{B}{}^{n+1}_{\epsilon_r}(x_j)$. Figure 10.3 shows the example of a segment which will be excluded.   □

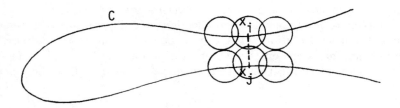

Fig. 10.3: Exclusion of a Pathological Approximating Segment

Finally, to advance the reader's intuition of an ϵ-near equilibrium path let us present three natural notions of *approximating equilibria for evolutions of economies.*

I. For the *exchange framework* we begin with a generalization of the well-known and fairly weak approximation criterion using the supremum norm.

Definition 10.6. *Let* $(\zeta_s)_{s\in[0,1]}$ *be an evolution of economies in one of the basic exchange models from Chapters 4 or 5, and let* $(\overline{p}, \overline{s})$ *be any point in the homotopy space* $X \times [0,1]$. *Then* $(\overline{p}, \overline{s})$ *is an* ϵ-*approximating equilibrium of the first kind for the evolution* $(\zeta_s)_{s\in[0,1]}$ *for some* $\epsilon > 0$ *if* $\max_{h=1,\dots,n} |\zeta_{h\overline{s}}(\overline{p})| < \epsilon$.

As Figure 10.4 below illustrates it an ϵ-approximating equilibrium of the first kind need not be close to a true equilibrium of the evolution.

II. The following notion applies to *any of our basic set-ups.*

Definition 10.7. *Let* $(\zeta_s)_{s\in[0,1]}$ *be an evolution of economies in any one of the basic models introduced in Part I, and let* $(\overline{p}, \overline{s})$ *be a point in the homotopy*

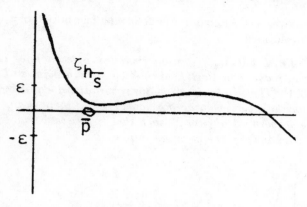

Fig. 10.4: Approximation Pathology of an ε-Near Equilibrium Path of the Second Kind I

space $X \times [0, 1]$. Then $(\bar{p}, \bar{s})$ is an **ϵ-approximating equilibrium of the second kind** for the evolution $(\zeta_s)_{s \in [0,1]}$ for an $\epsilon > 0$ if there is point (p_0, s_0) in the (relative) ϵ-neighborhood of $(\bar{p}, \bar{s})$ in $X \times [0, 1]$ which is a true equilibrium of the s_0-state economy ζ_{s_0} of the given evolution.

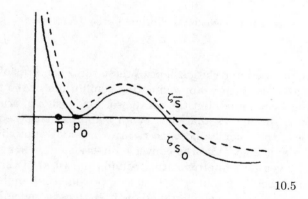

10.5

Fig. 10.5: Approximation Pathology of an ε-Near Equilibrium Path of the Second Kind II

Figure 10.5 illustrates the definition for the special case where $(\zeta_s)_{s \in [0,1]}$ is an evolution of exchange economies. Note that this notion still allows for shifts in the state-parameter in order to achieve a true equilibrium. Thus, an ϵ-approximating equilibrium of the second kind $(\bar{p}, \bar{s})$ still may be far distant from a true equilibrium of the $\bar{s}$-state economy $\zeta_{\bar{s}}$.

III. This, however, is excluded by the following last notion of approximating equilibria of evolutions.

Definition 10.8. *Let* $(\zeta_s)_{s\in[0,1]}$ *be an evolution of economies in any of the basic models introduced in Part I, and let* $(\overline{p}, \overline{s})$ *be a point in the homotopy space* $X \times [0,1]$*. Then* $(\overline{p}, \overline{s})$ *is an* **ϵ-approximating equilibrium of the third kind** *for the given evolution for some* $\epsilon > 0$ *if there is a point* p_0 *in the (relative)* ϵ*-neighborhood of* $\overline{p}$ *in* X *such that* $(p_0, \overline{s})$ *is a true equilibrium of the* $\overline{s}$*-state economy* $\zeta_{\overline{s}}$ *of the given evolution.*

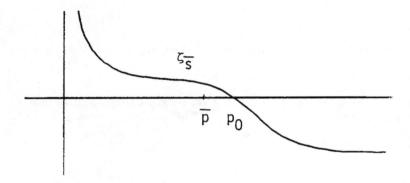

Fig. 10.6: ε-Near Equilibrium of the Third Kind

What are the precise interrelationships of these three concepts of approximating equilibria? Clearly, an ϵ-approximating equilibrium of the third kind is also one of the second kind, but the converse is obviously not true. Moreover, it is clear from the figures that neither an ϵ-approximating equilibrium of the third, nor of the second kind is necessarily also an ϵ-approximating equilibrium of the first kind. Also the converse directions do not generally hold. Nevertheless, given an ϵ-approximating equilibrium $(\overline{p}, \overline{s})$ of the third, or of the second kind there is for any $\epsilon' > 0$ an ϵ'-approximating equilibrium of the first kind in the relative ϵ-neighborhood of $(\overline{p}, \overline{s})$. This is immediate from the definitions.

Applying now these notions to the concept of an ϵ-near equilibrium path we arrive at the following conclusion:

Conclusion 10.9 *Any point on an* ϵ*-near equilibrium path for an evolution of economies is an* ϵ*-approximating equilibrium of the second kind, and under the hypothesis of Corollary 10.5 there is an* ϵ*-near equilibrium path all of whose points are even* ϵ*-approximating equilibria of the third kind.*

10.2 A Criterion for Identifying Points on Joining Equilibrium Components

The existence result of near-equilibrium paths naturally raises the general question of how to find near-equilibrium paths in the homotopy space for a given evolution. Apparently, this amounts to the general question of how one can discern for any given evolution whether a certain point in the homotopy space lies on one of its joining equilibrium components or not.

Here is a criterion which identifies *starting points* of joining equilibrium components. More formally speaking, we are going to provide a sufficient criterion for a point to be the starting point of a joining equilibrium component based on a local algebraic topological invariant. Moreover, we will see that this result also provides a partial answer to our general question.

Theorem 10.10. *Let $(\zeta_s)_{s \in [0,1]}$ be any evolution basing on any one of the basic models from Part I. Let furthermore $X \in \{\Delta^{n-1} \setminus L, \mathring{\Delta}^{n-1}, \mathbb{R}^n_+,$ $\mathbb{R}^n_+ \setminus (\bigcup_{i=1}^n D_i), \Delta^{n-1} \times [0, \beta], \Delta^{n-1} \times [0, \beta] \times [0,1], (G_-)^m \times (G_+)^m, C^{2m}\}$ denote the domain of the basic model. Choose an arbitrary connected component $A \subset X$ of the 0-state equilibrium set $\zeta_0^{-1}(0)$ of the given evolution. (For instance, A could be an isolated equilibrium vector of ζ_0.) Consider now that connected component R_A of the whole equilibrium set of the evolution which contains $A \times \{0\}$. Denote the intersection of R_A with the bottom $X \times \{0\}$ of the homotopy space by $\mathcal{A}$. (Notice that $\mathcal{A} \supseteq A$.)*

Then the following condition is sufficient for R_A to meet bottom and top of the homotopy space:

(*)   There is a neighborhood V_0 of $\mathcal{A}$ in $X \times \{0\}$ so that for any open neighborhood $\widetilde{W}$ of $\mathcal{A}$ in $X \times \{0\}$ which is contained in V_0 and has no equilibria of ζ_0 on its boundary $\overline{\widetilde{W}} \setminus \widetilde{W}$ the local fixed point index $i(g_0, \widetilde{W})$ of the equilibrium equivalent self-mapping of the 0-state economy is non-zero.

We will not give the details of the proof of Theorem 10.10. The reader can find them in Lehmann-Waffenschmidt (1985, pp. 33–34, Satz 2) for the special case $X = \mathring{\Delta}^{n-1}$. All steps of the proof given there immediately carry over to the general situation of Theorem 10.10. Here we confine ourselves to briefly commenting on the proof and then discussing the assumptions and the result of Theorem 10.10.

Theorem 10.10 follows from Theorem 3 by Browder (1960). However, Browder uses in his proof of Theorem 3 a stronger assumption than he requires in his formulation of the Theorem ('condition e'). Actually, one easily finds counterexamples against Browder's Theorem 3 as it is stated. Condition (*) in the formulation of our Theorem 10.10 is an adaption to our present situation of the stronger assumption actually used by Browder in his proof.

The following Figure 10.7 makes the hypothesis of Theorem 10.10 intuitive for the special case $X = \Delta^{n-1}$. All 'testing neighborhoods' $\widetilde{W}$ contain A and are contained in the 'reference neighborhood' V_0 of A.

Condition (∗) requires to compute the local fixed point index of g_0 for open neighborhoods $\widetilde{W}$. Naturally this requires explicit knowledge of the self-mapping g_0 to some extent. However, since we will not make here further use of the result of Theorem 10.10 we refer the interested reader to the extensive literature in algebraic topology on the topic of effective local index computation.

We are still left to demonstrate that Theorem 10.10 also gives at least a partial answer to the general question of how one can realise for an arbitrary point $(p, \bar{s})$ from the homotopy space $X \times [0, 1]$ whether it lies in a joining equilibrium component of some given evolution $(\zeta_s)_{s \in [0,1]}$, or not. Nevertheless,

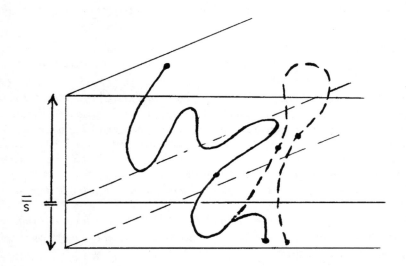

Fig. 10.7: Identifying Joining Equilibrium Components

this is straightforward. Just apply Theorem 10.10 twice, once to $(p, \bar{s})$ as a point of the domain of the initial state of the 'truncated' evolution $(\zeta_s)_{s \in [\bar{s},1]}$, and once to $(p, \bar{s})$ as a point of the domain of the initial state of the remaining part $(\zeta_{\bar{s}-\sigma})_{\sigma \in [0,\bar{s}]}$ of $(\zeta_s)_{s \in [0,1]}$ which is reversely parametrized. When both times the criterion of Theorem 10.10 is satisfied, then $(p, \bar{s})$ clearly lies on a joining equilibrium component of the whole evolution $(\zeta_s)_{s \in [0,1]}$.

Notice, however, that this method only identifies points $(p, \bar{s})$ on such joining equilibrium components which are separated by the $\bar{s}$-slice of the homotopy space into two connected parts completely lying below and above the $\bar{s}$-slice respectively. This means, the dotted case in Figure 10.7 for example is not within the scope of this method.

11

Equilibrium Paths

Time has now come to fulfill our promise from Chapter 10. There we have resorted to geometrical intuition arguing that geometrically bad parts of a joining equilibrium component are always on account of a more or less "pathological" variation of economic states. Unfortunately the term "pathological" only can serve as an intuitive guide, not as a firm base of further analysis. What we will do now is to show precisely that intuition actually led us the right way.

Obviously one possible way to do that would be to classify all "bad parts" which may appear in joining equilibrium components and then to show that they are always generated by pathologically looking evolutions. Unfortunately, however, this program would not only mean a hard piece of technical work. Still worse, it is also grounded on an unsound base because it essentially relies on the reseracher's subjective opinion as to which evolutions should be judged as pathological.

Thus, we will adopt another way – starting from the opposite direction. To be more specific we will show that for any of our basic models there is a large class of evolutions which are highly well-behaved and reasonable from the mathematical and from the economic viewpoint as well, and which, particularly, have nicely behaved joining equilibrium paths. This especially means that none of the geometrically bad features discussed in Section 10.1 above can occur. Moreover, we will also be able to show that it is even possible to approximate any evolution arbitrarily closely by such a well-behaved evolution.

Our specific procedure will be first to provide a thorough analysis of these issues in Section 11.1 for our reference basic model of pure exchange of the Dierker type. Actually, we will provide *two different classes* of well-behaved and approximating exchange evolutions in Section 11.1. The first one is essentially based on the method of polynomial approximation which is fairly popular in economies. The second one is based on an approximation method which is even simpler than polynomial approximation and which has the further advantage of being entirely *constructive*.

In Section 11.2 then we will point out the necessary modifications to generalize the two developed methods to the other basic models. In the last Section 11.3 of this Chapter we demonstrate that these methods immediately yield several *refined connectedness results* on the graphs of the equilibrium correspondences of our basic models. The main justification for the attribute "refined" will be given in Section 14.1 where we will provide a comparison with Balasko's well-known connectedness results on the graph of the Walras correspondence.

Let us conclude this Introduction by the following remark. There is also a close relationship of our results given in this Section to a result by Mas-Colell given in his book (1985, Section 8.8). To be more specific Mas-Colell shows for the Walrasian exchange model that the subspace of so-called "regular one-parametrizations" is open dense in the space of one-parametrizations (i.e. exchange evolutions). Particularly, regular one-parametrizations have rather well-behaved joining equilibrium components. Unfortunately, however, we will see that unlike our well-behaved joining equilibrium paths the well-behaved joining equilibrium components in Mas-Colell's sense still allow for some of the geometrically bad features which we have discussed in Section 10.1 for path-connected joining equilibrium components.

Nevertheless, Mas-Colell's result, remarkable as it is, only gives existence of approximating one-parametrizations with well-behaved joining equilibrium congruents. It does not provide any constructive method how to achieve them. In this Chapter we will go beyond this abstract existence result in that we explicitly provide methods to construct approximating evolutions with nicely one-parametrizable joining equilibrium components. In Section 14.2 the reader can find a comprehensive discussion on the relationship of our approach and results to those of the differentiable approach employed by Mas-Colell.

11.1 Approximating Evolutions of Exchange Economies With Nice Equilibrium Paths Based on Dierker's Model from Section 4.2

We start our investigations with the basic model of the Dierker type. We do so since this model notably well suits as a reference model for later generalizations.

We are going to present *two intuitive general methods* for approximating any given Walrasian exchange evolution by a well – behaved and economically intuitive evolution of this type having, moreover, nicely behaved joining equilibrium paths. Our *first method* (Subsection 11.1.1), grounded on *approximation by polynomials*, has the advantage that the construction of a well – behaved approximating evolution is straightforward. To prove the well – behavedness of the equilibrium set, however, requires an advanced result from algebraic geometry. Even though, this method is far more *constructive* than the addressed abstract existence result by Mas-Colell. The reader should note, however, that the attribute constructiveness can, to be honest, only be credited to it as far as the current state of mathematical approximation theory justifies it. This reservation does not apply to our *second method*, which, however, is a little bit more laborious. It achieves approximating evolutions in a completely constructive manner using as main analytical tool piecewise linear functions. This construction has the further advantage that it makes the well – behavedness of the equilibrium set still more intuitive.

11.1.1 Polynomial Approximating Exchange Evolutions

Let an evolution $\zeta = (\zeta_{i_s})_{s \in [0,1]}^{i=1\ldots n}$ basing on Dierker's version of a Walrasian exchange economy and any $\epsilon > 0$ be given. Now we have a twofold purpose. First we will set about providing a general and intuitive construction of a Walrasian exchange evolution which ϵ-*approximates* the given one, and then we have to convince ourselves that our achieved evolution indeed possesses a "nicely behaved" joining equilibrium component, i.e. a nicely behaved equilibrium path.

Before we can set to work constructing a suitable ϵ-approximating evolution, however, we clearly first have to render precise the notion of a "nicely behaved joining equilibrium component". Nevertheless, from the discussion in Chapter 10 it is intuitively clear what we expect from a joining equilibrium component deserving this attribute. In a word it should be *well passable* in the *intuitive geometrical sense* from bottom to top of the homotopy price prism. To call it to the reader's mind this particularly means that it contains at least one joining path consisting exclusively of equilibria which, first and foremost, is continuous and of finite Euclidean length, and, moreover, does nowhere perform infinitely many oscillations.

But how can we formalize this in a precise analytical way? Actually, it is not hard to specify *certain* unwelcome characteristics of continuous paths as we have just done, or to rigorously eliminate *any* complications by confining ourselves to the very restricted class of finitely piecewise linear paths, as we did in Chapter 10. On the other hand, it is not at all clear from the outset how to design the proper class of all *"well passable" continuous paths* in a comprehensive and comprehensible way. Moreover, difficulties are still increased by the fact that any such design clearly must be intimately linked to our later construction of ϵ-approximating evolutions. Actually, however diligent we may design a class of all well passable continuous paths, it is surely not much use if we later cannot make out the joining equilibrium components of our ϵ-approximating evolutions falling into this class.

Fortunately, we can present a class which serves all of our purposes. From our construction the reader will immediately see that it is built on a geometrically intuitive principle. To start now we call a *path* in Euclidean space

$$w : [a, b] \longrightarrow \mathbb{R}^n$$

(real) analytical if there exist real numbers $c < a$ and $d > b$ and n analytical functions

$$\widetilde{w}_i :]c, d[\longrightarrow \mathbb{R}, \quad i = 1, \ldots, n,$$

such that each $\widetilde{w}_i$ extends the i–th component function w_i of w, i.e.

$$\forall_{i=1,\ldots,n} \ \widetilde{w}_i|_{[a,b]} \equiv w_i.$$

Remember that the *arc* of a path w is its image $w([a,b]) \in \mathbb{R}^n$. This means that the path w represents the *one-parametrization* of the arc. Or, in other words, the arc shows the geometric configuration shaped by the moving parameter $w(t)$, $t \in [a,b]$. Any of the subsequent concepts and results in this Section are valid for general paths which, in particular, may be non one-to-one maps, which means that they may "intersect themselves", or in other words, that the path essentially differs from its arc. However, in this study, paths are usually one-to-one maps, in accordance with the purpose of modelling and analyzing "passability". In fact, analytical paths are geometrically remarkably well–behaved curves. This will be made lucid by the characterizations provided by Proposition 11.1 below. Proposition 11.1 has both virtues of being intuitive and precise. It makes essential use of what we call a *normalized tangent direction*. More specifically, let us denote the derivative function of an analytical path

$$w : [a, b] \longrightarrow \mathbb{R}^n$$

by

$$w' = (w_1', \ldots, w_n') : [a, b] \longrightarrow \mathbb{R}^n.$$

Furthermore, as suggested by geometrical intuition, for any $t_0 \in [a,b]$ with $w'(t_o) \neq 0^n$ we call the straight line in $\mathbb{R}^n$

$$w(t_0) + \alpha \cdot w'(t_0), \ \alpha \in \mathbb{R},$$

the *tangent line* at $w(t_0)$ (see Figure 11.1).

Furthermore, we call any vector of the form $\alpha \cdot w'(t_o), \alpha \in \mathbb{R} \backslash \{0\}$, a *tangent direction at the point* $w(t_o)$ corresponding to t_o. (Note that possibly $w(t_o) = w(t_1)$ for $t_1 \neq t_o$.) Evidently the tangent direction is only determined up to a multiple. Nevertheless, under the additional assumption that $w_1'(t_o) \neq 0$ it is possible to make a *unique* choice $v = (v_1, \ldots, v_n)$ of the tangent direction by normalizing it such that $v_1 = 1$. Accordingly,

$$v(t_o) = \frac{1}{w_1'(t_o)} \cdot w'(t_o).$$

Actually, the normalized tangent direction gives us an intuitive and exact instrument to study the behaviour of an analytical path in dependence of the parameter t. This is realized by Proposition 11.1 below which demonstrates that analytical paths particularly cannot oscillate infinitely often. Note in particular that for any $t \in]a_i, a_{i+1}[, i = 0, \ldots, r - 1$, the vector

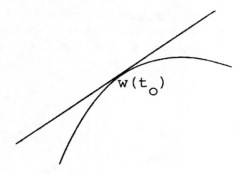

Fig. 11.1: Tangent Line

$$\left(1, \frac{w_2'}{w_1'}(t), \ldots, \frac{w_n'}{w_1'}(t)\right) \quad \text{with} \quad \frac{w_j'}{w_1'}(t) = \frac{w_j'(t)}{w_1'(t)}$$

is a normalized tangent direction.

Proposition 11.1. *Let any analytical path* $w' : [a,b] \longrightarrow \mathbb{R}^n$ *be given whose component functions* $w_1', \ldots, w_n'$ *are all non-constant. Then there exists a finite subdivision*

$$a = a_0 < a_1 < \ldots < a_r = b \quad \text{of the interval} \quad [a,b]$$

such that the following properties hold for all $j = 1, \ldots, n$ *and* $i = 0, \ldots, r-1$

(i) w_j' *is strictly monotonic on each closed subinterval* $[a_i, a_{i+1}]$,
(ii) $w_1'(t) \neq 0$ *on each open subinterval* $]a_i, a_{i+1}[$,
(iii) w_j'/w_1' *is constant, or strictly monotonically increasing, or decreasing on each open subinterval* $]a_i, a_{i+1}[$.

Proof. The Proposition is an immediate Corollary to the following standard result on analytical functions:

Proposition 11.2. *Let* $h_1, h_2 :]c, d[\longrightarrow \mathbb{R}$ *be analytical functions such that* h_2 *has no zeroes in* $]c, d[$. *Then*

(i) $h_1 \equiv 0$, *or* h_1 *has at most finitely many zeroes in any compact subinterval* $[a, b] \subset]c, d[$.
(ii) h' *and* h_1/h_2 *are analytical where* h_1' *denotes the derivative of* h_1.

Some remarks on Proposition 11.1 seem worthwhile. Clearly, the assumption that no component function of w' is constant obviously is not restrictive. In fact, a constant component function w_j' only reduces the problem by one dimension, since $w' : [a,b] \longrightarrow \mathbb{R}^{n-1} \times \{pt\}$.

To illuminate Proposition 11.1 further let us have a look at the following examples of Figures 11.2a and 11.2b showing two analytical paths with a convex and a concave arc respectively. The reader should particularly note that in both examples *both components* w_1, w_2 of the analytical path w : $[a, b] \longrightarrow \mathbb{R}^2$ are *strictly monotonically increasing.*

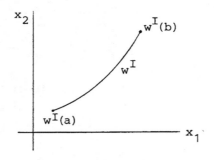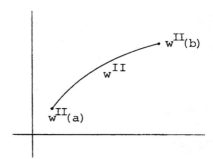

(a) $\frac{w_2'}{w_1'}$ Strictly Monotonically Increasing Curve

(b) $\frac{w_2'}{w_1'}$ Strictly Monotonically Decreasing Curve

Fig. 11.2

To *sum up*, Proposition 11.1 guarantees that the arc of an analytical path consists of finitely many parts which are either linear segments, or which essentially look like the highly regular arcs of Figures 11.2a and 11.2b.

Actually, the notion of an analytical path leads us close to our aim of designing the class of all "well passable" paths. However, the following example exhibits that we are still not quite done. Consider the subspace of $\mathbb{R}^2$ formed by the two coordinate axes. Clearly, there exists an analytical path joining $(1, 0)$ and $(0, 0)$, and one joining $(0, 0)$ and $(0, 1)$. But, there is no analytical path joining $(1, 0)$ and $(0, 1)$! This follows immediately from the following

Proposition 11.3. *Let* $w : [a, b] \longrightarrow \mathbb{R}^n$ *be an analytical path and* $f : \mathbb{R}^n \longrightarrow \mathbb{R}$ *be any polynomial from* $\mathbb{R}[x_1, \ldots, x_n]$. *Then either* $w([a, b]) \subset f^{-1}(0)$, *or the set* $w^{-1}(f^{-1}(0))$ *is finite. (In words the latter means that the arc of* w *intersects the zero set of* f *in at most finitely many points.)*

Proof.   Proposition 11.3 is also an immediate Corollary of Proposition 11.2.

Designing a class of "well passable" paths which rules out right angles clearly would be highly inadequate. Nevertheless, there is a simple way to cure our approach from this unwelcome oversharpness. We just admit finite compositions of analytical paths, i.e. "finitely piecewise analytical paths". More precisely, we define:

Definition 11.4. *A continuous path* $w : [a, b] \longrightarrow \mathbb{R}^n$ *is called a **(finitely) piecewise analytical path** if there exists a finite subdivision*

$$a = a_0 < a_1 < \ldots < a_m = b$$

such that all restrictions $w_{[a_i,a_{i+1}]}$ are analytical paths. (Thus, a (finitely) piecewise analytical path may have finitely many undifferentiable kinks.) Let us call a subset $A \subset \mathbb{R}^n$ finitely piecewise analytically path connected if any two points of A can be joined by a (finitely) piecewise analytical path which lies entirely in A. A (finitely) piecewise analytical path component of some space B accordingly is a subset of B which is maximal with this property.

The following Figure 11.3 shows an example of a piecewise analytically path connected set.

Fig. 11.3: Piecewise Analytically Path Connected Set

Actually, the class of finitely piecewise analytical paths is our desired class of well passable paths. A finitely piecewise analytical path is intuitively well passable, and over and above that, it seems hard to conceive of intuitively well behaved paths which are not from this class.

Now we are prepared to tackle our main task of this Section. That means we are going to show that the method of polynomial approximation which is fairly popular in economics is also useful for our present purposes. More specifically, employing the polynomial approximation method we will first provide an evolution $\overline{\zeta}$ which ϵ–approximates the initially given evolution ζ, and then we will demonstrate that the equilibrium set of $\overline{\zeta}$ actually contains a joining piecewise analytical equilibrium path.

At the outset of our construction let us choose a positive $\delta < 1/n$ with the property that for any $p \in \mathring{\Delta}^{n-1} \backslash \Delta_\delta^{n-1}$ with $p_i < \delta$ and any $s \in [0,1]$ one has $\zeta_{i_s}(p) > 2$. Accordingly, the equilibrium set of ζ is contained in $\Delta_\delta^{n-1} \times [0,1]$. (The reason for choosing 2 as bounding value will become clear during the construction.) Our construction is essentially based on the famous Weierstraß

Approximation Theorem (see e.g. Lang, 1969, p. 152, Corollary). Without loss of generality let us assume that $0 < \epsilon < 1$. Now for each of the first $n-1$ restricted market evolutions

$$(\zeta_{i_s})_{s\in[0,1]}\,|_{\Delta_{\delta/2}^{n-1}\times[0,1]}, \quad i = 1,\dots,n-1,$$

of the given evolution we choose a polynomial

$$Q_i : \mathbb{R}^n \times \mathbb{R} \longrightarrow \mathbb{R}$$

which η–approximates ζ_i uniformly on $\Delta_{\delta/2}^{n-1} \times [0,1]$ with $\eta := \frac{\delta \cdot \epsilon}{4 \cdot n}$. Due to Walras' law one has the following equality for any $p \in \Delta_{\delta/2}^{n-1}$ and $s \in [0,1]$

$$\zeta_{n_s}(p) = -\frac{1}{p_n} \sum_{i=1}^{n-1} p_i \cdot \zeta_{i_s}(p).$$

Consequently the function

$$Q_n : \Delta_{\delta/2}^{n-1} \times [0,1] \longrightarrow \mathbb{R} \quad \text{with}$$

$$Q_n(p,s) := -\frac{1}{p_n} \sum_{i=1}^{n-1} p_i \cdot Q_i(p,s)$$

$[2(n-1) \cdot \eta/\delta]$–approximates the n-th component function ζ_n uniformly on $\Delta_{\delta/2}^{n-1} \times [0,1]$. (Clearly the function Q_n is *not a polynomial*, but this does not matter for our purposes.) Let

$$Q : \Delta_{\delta/2}^{n-1} \times [0,1] \longrightarrow \mathbb{R}^n$$

be the function with components $Q_1,\dots,Q_n$. By construction it satisfies Walras' law. Furthermore, due to the choice of η we get

$$\|\zeta_s(p) - Q(p,s)\| < \epsilon \quad \text{for all} \ (p,s) \in \Delta_{\delta/2}^{n-1} \times [0,1].$$

Now let us choose some continuous gluing function

$$\alpha : \mathring{\Delta}^{n-1} \longrightarrow [0,1]$$
$$\text{with} \quad \alpha|_{\Delta_\delta^{n-1}} \equiv 0, \quad \text{and}$$
$$\alpha|_{\mathring{\Delta}^{n-1}\setminus\mathring{\Delta}_{\delta/2}^{n-1}} \equiv 1.$$

We define the continuously one–parametrized family

$$\overline{\zeta} = (\overline{\zeta}_{i_s})_{s\in[0,1]}^{i=1} : \mathring{\Delta}^{n-1} \times [0,1] \longrightarrow \mathbb{R}^n$$
$$(p,s) \longmapsto \alpha(p)\zeta_s(p) + (1 - \alpha(p))Q(p,s).$$

Thus we obtain a family of mappings which on the inner part $\Delta_\delta^{n-1} \times [0,1]$ of the price prism equals Q and on the $\delta/2$–neighborhood of the boundary equals the given evolution ζ. On the area in between, i.e. on $\Delta_{\delta/2}^{n-1} \backslash \Delta_\delta^{n-1}$, Q is continuously transformed into ζ.

From the construction it is immediately clear that $\bar{\zeta}$ is an evolution which ϵ-approximates the given evolution ζ uniformly on the whole price space $\mathring{\Delta}^{n-1} \times [0,1]$.

Thus it remains to us to verify

Proposition 11.5. $\bar{\zeta}$ *possesses at least one joining piecewise analytical equilibrium path.*

Proof. Let us choose some joining equilibrium component C of $\bar{\zeta}$ according to Theorem 10.2. Note that due to Walras' law C is contained in the set

$$\bigcap_{i=1}^{n-1} \left(\left(\bar{\zeta}_{i_s} \right)_{s \in [0,1]} \right)^{-1} (0).$$

If we can show that this intersection set is semi–algebraic (see the "Mathematical Preliminaries") then the following Proposition 11.6 tells us that C is finitely piecewise analytically path connected, and we are done.

Proposition 11.6. *Let $A \subset \mathbb{R}^n$ be a semi–algebraic set. Then any connected component Z of A is even finitely analytically path connected.*

We will postpone the proof of Proposition 11.6 to the end of the main line of our proof of Proposition 11.5.

To see that the above intersection set actually is semi–algebraic is not hard. Note that by construction it equals the set

$$\{(p,s) \in \mathbb{R}^n \times \mathbb{R} | Q_1(p,s) = \ldots = Q_{n-1}(p,s) = 0\} \cap (\Delta_\delta^{n-1} \times [0,1]).$$

It is easy to see that this intersection is semi–algebraic: the first set is semi–algebraic by property (i) of semi–algebraic sets (see "Mathematical Preliminaries"). $\Delta_\delta^{n-1} \times [0,1]$ is semi–algebraic by properties (ii) and (iii), and the intersection again is semi–algebraic by property (iii).

The reader should note that our argumentation not only shows that $\bar{\zeta}$ possesses a joining finitely piecewise analytical equilibrium path, but even that *any connected component* of the equilibrium set of $\bar{\zeta}$ is finitely piecewise analytically path connected.

Finally we are left to prove Proposition 11.6.

Proof of Proposition 11.6 The proof is essentially based on the following deep mathematical result by B. Teissier (1975, Prop. 3, p. 313) which, nevertheless, will be easy to understand for the reader:

Theorem 11.7. *Given a semi–algebraic subset $A \subset \mathbb{R}^n$ there exists for any $y \in A$ an arbitrarily small open neighborhood $U(y) \subset \mathbb{R}^n$ such that $U(y) \cap A$ is piecewise analytically path connected.*

To start the main line of the proof of Proposition 11.6 let us choose a *connected component* $Z \subset A$ and an arbitrary point $x \in Z$. We define the following subset of Z

$$Z(x) := \{z \in Z| \text{ there exists a piecewise analytical path} \\ \text{in } Z \text{ connecting } x \text{ and } z\}.$$

(This means that $Z(x)$ is the piecewise analytical path component of Z which contains x.)

Clearly we are done when we can show that $Z = Z(x)$.

Let us start with recalling the following elementary facts about Z and $Z(x)$:

(1) $Z \subset A$ is an open and closed subset.

(2) If $Y \subset Z$ is non–empty and open *and* closed relatively to Z, then $Y = Z$.

(3) If $w : [a, b] \longrightarrow A$ is a continuous path which meets Z, then actually $w : [a, b] \longrightarrow Z$.

(4) If $w : [a, b] \longrightarrow A$ is a piecewise analytical path which meets $Z(x)$, then actually $w : [a, b] \longrightarrow Z(x)$.

Due to (2) it suffices to prove that $Z(x)$ is a relatively *open and closed* subset of Z.

(i) $Z(x) \subset Z$ is *open* in Z : choose an arbitrary $y \in Z(x)$ and an open neighbourhood $U(y) \subset \mathbb{R}^n$ as in the result by B. Teissier cited above. From (4) follows that $U(y) \cap A \subset Z(x)$. This proves the assertion.

(ii) $Z(x) \subset Z$ is *closed* in Z, i.e. $Z \backslash Z(x) \subset Z$ is open: let $y \in Z \backslash Z(x)$ and choose an open neighborhood $U(y) \subset \mathbb{R}^n$ as in (i) such that $U(y) \cap A$ is piecewise analytically path connected. We prove that $U(y) \cap A \subset Z \backslash Z(x)$. By (3) above $U(y) \cap A$ is contained in Z. Assume now that there exists a point $z \in U(y) \cap Z(x)$. Then it is clear from the construction that there are two piecewise analytical paths in Z connecting z with y and z with x respectively. But then $y \in Z(x)$, and this is a contradiction. Therefore $U(y) \cap A \subset Z \backslash Z(x)$. Since this is true for all y, the set $Z \backslash Z(x)$ is open in Z, and Proposition 11.6 is proved. $\square$

11.1.2 Piecewise Linear Approximating Exchange Evolutions

In the preceding Section we have employed the popular polynomial approximation method in order to achieve approximating exchange evolutions with nicely behaved joining equilibrium components. This method clearly is more constructive than the pure existence result by Mas-Colell addressed in the Introduction to this Chapter and in the General Introduction. Nevertheless, our method clearly can be viewed only as constructive as far as the current state of mathematical approximation theory justifies it.

Thus the natural question remains whether it is possible to provide a *fully constructive method* serving our purposes. We will see in this Section that this is possible. Actually, the method we are going to present fulfills both requirements of being intuitive from the economic as well as from the mathematical viewpoint. Comparing this method to the preceding polynomial method the reader may find that the construction of an approximating evolution is more laborious, even though it is straightforward. However, in our eyes, this is more than compensated by the fact that it is already clear from the construction that joining equilibrium components are well–behaved. The latter can also be obtained as a special case from Proposition 11.6 above.

Our basic idea to achieve a fully constructive method of approximating any Walrasian evolution by one with a nicely behaved equilibrium set is to exclusively use convex transitions of excess demand functions whose graphs over Δ_ϵ^{n-1} consist of finitely many simplices. It will be intuitively clear that choosing the simplices small enough and taking sufficiently many convex transitions achieves any desired approximation quality. Let us now see the details.

Let any evolution $\zeta = (\zeta_{i_s})_{\substack{i=1 \\ s\in[0,1]}}^{n}$ basing on Dierker's version of a Walrasian exchange economy and any $\epsilon > 0$ be given. As before we aim at providing an evolution $\widetilde{\zeta} = (\widetilde{\zeta}_{i_s})_{\substack{i=1 \\ s\in[0,1]}}^{n}$ from this class which ϵ–approximates ζ on the whole domain $\mathring{\Delta}^{n-1} \times [0,1]$ and which has joining finitely piecewise analytical equilibrium paths.

Let us begin with a brief outline of our construction. As before let us choose a $\delta > 0$, such that for any $p \in \mathring{\Delta}^{n-1}\backslash\Delta_\delta^{n-1}$ with $p_i < \delta$ one has $\zeta_{i_s}(p) > 2$ for any $s \in [0,1]$. Thus, the equilibrium set of ζ is contained in $\Delta_\delta^{n-1} \times [0,1]$. Now we focus on the s–state economy $\zeta_s = (\zeta_{i_s})_{i=1}^{n}$ for some arbitrarily chosen $s \in [0,1]$. Actually, instead of approximating the *whole* evolution ζ on $\Delta_{\delta/2}^{n-1} \times [0,1]$ as we did before, we now first approximate the s–state economy ζ_s on $\Delta_{\delta/2}^{n-1}$. Then, like previously, we will glue the approximating component functions $\widetilde{\zeta}_{i_s}$ with the given component functions ζ_{i_s} over the area $\Delta_{\delta/2}^{n-1}\backslash\Delta_\delta^{n-1}$ by means of the convex transition. In a final step then we will convince ourselves that finitely many convex connection economies between such approximating state economies actually are enough to achieve an evolution with the desired properties.

Let us now see how the construction works. We start with $\epsilon/2$–approximating the first $n-1$ restricted component functions $\zeta_{i_s}|_{\Delta^{n-1}_{\delta/2}}$, $i = 1, \ldots, n-1$, by means of *piecewise linear functions*. (The reason for choosing $\epsilon/2$ will become clear from our later construction.) To this end we choose for each $i = 1, \ldots, n-1$ a *finite triangulation* $\sum_{i_s}$ of $\Delta^{n-1}_{\delta/2}$, i.e. a decomposition of $\Delta^{n-1}_{\delta/2}$ into finitely many $(n-1)$–simplices $\sigma_1, \ldots, \sigma_{k_{i_s}}$, such that the following holds: for $i = 1, \ldots, n-1$ consider any decomposing domain simplices $\sigma_j \in \sum_{i_s}$ and the convex hull $\overline{\sigma}_j$ in $\Delta^{n-1}_\delta \times \mathbb{R}$ of the n points $(v_{r_j}, \zeta_{i_s}(v_{r_j}))$ where v_{r_j} are the vertices of σ_j. Loosely speaking, $\overline{\sigma}_j$ is the $(n-1)$–simplex in $\Delta^{n-1}_\delta \times \mathbb{R} \subset \mathbb{R}^{(n-1)+1}$ whose vertices are the images under $\zeta_{i_s}|_{\Delta^{n-1}_\delta}$ of the vertices of the domain simplex σ_j. Clearly, it is possible to choose the triangulation $\sum_{i_s}$ fine enough so that any $\overline{\sigma}_j, j = 1, \ldots, k_{i_s}$, $\eta/2$–approximates the graph $\zeta_{i_s}(\sigma_j)$ of ζ_{i_s} over σ_j with $\eta := \frac{\delta \cdot \epsilon}{4n}$. Now extend the triangulations $\sum_{i_s}, i = 1, \ldots, n-1$, from Δ^{n-1}_δ to $\Delta^{n-1}_{\delta/2}$. Let us denote the obtained triangulation of $\Delta^{n-1}_{\delta/2}$ by $\sum'_{i_s}$, and its restriction to the area $\Delta^{n-1}_{\delta/2} \backslash \mathring{\Delta}^{n-1}_\delta$ by $\sum''_{i_s}$. Again let us choose the triangulation $\sum''_{i_s}$ fine enough so that we arrive at a function defined on $\Delta^{n-1}_{\delta/2}$ whose graph consists of $k_{i_s} + k'_{i_s}$ $(n-1)$–dimensional simplices, i.e. which is *finitely piecewise linear*, and which $\eta/2$–approximates ζ_{i_s} on $\Delta^{n-1}_{\delta/2}$.

Now let us choose a finite triangulation $\sum'_s = \{\sigma'_1, \ldots, \sigma'_2\}$ of $\Delta^{n-1}_{\delta/2}$ which *refines* the triangulations $\sum'_{1_s}, \ldots, \sum'_{n-1_s}$. Denote the restriction of $\sum'_s$ to $\Delta^{n-1}_{\delta/2} \backslash \mathring{\Delta}^{n-1}_\delta$ by $\sum''_s$, and for each $i = 1, \ldots, n-1$ denote by $\widetilde{\zeta}_{i_s}|_{\Delta^{n-1}_{\delta/2}}$ the finitely piecewise linear function which is constructed on the above principles using the triangulation $\sum'_s$. As before let us define

$$\widetilde{\zeta}_{n_s}|_{\Delta^{n-1}_{\delta/2}}(p) = -\frac{1}{p_n} \sum_{i=1}^{n-1} \widetilde{\zeta}_{i_s}(p) \cdot p_i.$$

($\widetilde{\zeta}_{n_s}|_{\Delta^{n-1}_{\delta/2}} ((n-1) \cdot \eta/\delta)$–approximates the given $\zeta_{n_s}|_{\Delta^{n-1}_{\delta/2}}$.) It follows immediately from the construction that $\widetilde{\zeta}_s|_{\Delta^{n-1}_{\delta/2}} = (\widetilde{\zeta}_{i_s}|_{\Delta^{n-1}_{\delta/2}})_{i=1}^n$ $\epsilon/2$–approximates ζ_s on $\Delta^{n-1}_{\delta/2}$.

In the next step we glue $\widetilde{\zeta}_s|_{\Delta^{n-1}_{\delta/2}}$ together with ζ over the area $\overline{\Delta}^{n-1}_{\delta/2} \backslash \Delta^{n-1}_\delta$ in the same manner as before using the gluing function $\alpha : \mathring{\Delta}^{n-1} \longrightarrow [0, 1]$ with

$$\alpha|_{\Delta^{n-1}_\delta} \equiv 0 \quad \text{and}$$

$$\alpha|_{\mathring{\Delta}^{n-1}\setminus\mathring{\Delta}^{n-1}_{\delta/2}} \equiv 1.$$

Thus, we arrive at the function

$$\widetilde{\zeta}_s : \mathring{\Delta}^{n-1} \longrightarrow \mathbb{R}^n$$

$$p \mapsto \begin{cases} \widetilde{\zeta}_s|_{\Delta^{n-1}_{\delta/2}}(p) & \text{for } p \in \Delta^{n-1}_\delta \\ \alpha(p)\zeta_s(p) + (1-\alpha(p))\widetilde{\zeta}_s(p) & \text{for } p \in \Delta^{n-1}_{\delta/2}\setminus\Delta^{n-1}_\delta \\ \zeta_s(p) & \text{for } p \in \mathring{\Delta}^{n-1}\setminus\Delta^{n-1}_{\delta/2} \end{cases}$$

Clearly, $\widetilde{\zeta}_s$ is an economy which $\epsilon/2$–approximates the given s–state economy ζ_s. Due to its properties we will call $\widetilde{\zeta}_s$ ($\widetilde{\zeta}_{i_s}$) in the sequel a *finitely piecewise linear $\epsilon/2$–approximating s–state economy (market)*.

Let us now look at the whole evolution ζ. Clearly it would be both economically and geometrically appealing to construct an approximating evolution by exclusively using convex transitions of finitely piecewise linear ϵ–approximating state economies. Actually, this will be our approach.

First let us fix any $i \in \{1, \ldots, n-1\}$ and any s–state market $\zeta_{i_s}, s \in [0,1]$. The following simple observation provides the key for our later construction. Due to the construction of $\widetilde{\zeta}_s$ and the allover uniform continuity of ζ there is a positive real number ϑ_s such that $\widetilde{\zeta}_{i_s}$ $\epsilon/2$–approximates $\zeta_{i_{s'}}$ for all $s' \in]s-\vartheta_s, s+\vartheta_s[\cap [0,1]$. Now, the remainder of our construction will essentially be an exploitation of the compactness of $[0,1]$, and a piece of handcraft.

Let us start with choosing a finite relatively open subcovering $\{[0, \vartheta_{i_0}, [,]s_1^i - \vartheta_{i_1^s}, s_1^i + \vartheta_{s_1^i}[, \ldots,]s_l^i - \vartheta_{s_l^i}, s_l^i + \vartheta_{s_l^i}[,]1 - \vartheta_{i_{l+1}}, 1]\}$ of $[0,1]$. Now let us focus on the *overlapping open intervals* $]s_{j+1}^i - \vartheta_{s_{j+1}^i}, s_j^i + \vartheta_{s_j^i}[$, $j = 0, \ldots, l$, where we take the convention that $s_0^i := 0$ and $s_{l+1}^i := 1$. Denote furthermore the midpoint of any overlapping interval $]s_{j+1}^i - v_{s_{j+1}^i}, s_j^i + v_{s_j^i}[$ by S_j^i. Hence

$$S_j^i = s_j^i + v_{s_j^i} - 1/2(s_j^i + v_{s_j^i} - (s_{j+1}^i - v_{s_{j+1}^i}))$$
$$= s_{j+1}^i - v_{s_{j+1}^i} + 1/2(s_j^i + v_{s_j^i} - (s_{j+1}^i - v_{s_{j+1}^i})),$$

$j = 0, \ldots, l$, lies in $[S_j^i, S_{j+1}^i]$, but generally does not equal the midpoint $\frac{S_j^i + S_{j+1}^i}{2}$ of this interval. The following Figure 11.4 summarizes that. Thus we have $(2l + 3)$ distinguished points $s_0^i = 0 < S_0^i < s_1^i < S_1^i < s_2^i < S_2^i < \ldots < s_l^i < S_l^i < s_{l+1}^i = 1$ forming a partition of $[0,1]$. What we are heading for is the *approximating market evolution* composed of the $2l+2$ convex transitions between any two approximating economies with neighboring indices from this ordered list (symbolized by the dotted lines in Figure 11.4). However, in view of our ultimate aims there is the last difficulty that the triangulations $\sum'$ of $\Delta^{n-1}_{\delta/2}$ associated with each of these points according to our previous considerations clearly need not coincide. But we can easily overcome this problem

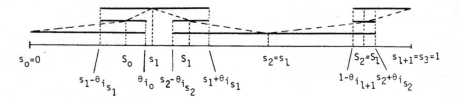

Fig. 11.4: Construction Scheme for an Approximating Market Evolution

by just taking a common refinement $\sum_i$ of these $2l + 3$ triangulations. This means we now form the $2l + 3$ finitely piecewise linear $\eta/2$–approximating markets $\widetilde{\zeta}_{i(-)}$ for commodity i at the $2l + 3$ points of the above list on the basis of the refined triangulation $\sum_i$. Thus we are in a position to form the $(2l + 2)$–fold composite market evolution $\hat{\zeta}_i$ composed of the $2l + 2$ convex connection market evolutions between any two $\eta/2$–approximating markets with neighboring indices from the above ordered list. The reader can easily convince himself that $\hat{\zeta}_i$ actually $\eta/2$–approximates the given market evolution ζ_i. Furthermore, there are simple examples showing that this is no longer true if one would omit the "overlapping midpoints" S_j from the construction.

Now the last steps of our construction are as follows. First we have to perform the outlined procedure for each of the first $n - 1$ market evolutions ζ_i, $i = 1, \ldots, n - 1$, using a common refinement $\sum$ of the triangulations $\sum'_i$, $i = 1, \ldots,$
$n - 1$. Then we take the common refinement R of the $n - 1$ partitions $s_0^i = 0 < S_0^i < s_1^i < S_1^i < \ldots < s_l^i < S_l^i < s_{l+1}^i = 1, i = 1, \ldots, n - 1$, and again have to re–perform our procedure for the first $n - 1$ market evolutions ζ_i, $i = 1, \ldots, n - 1$, with the refined partition R of $[0, 1]$. Using the Walras formula as before we achieve an $n - th$ market evolution $\hat{\zeta}_n$ which $\frac{(n-1)\cdot\eta}{\delta}$–approximates the given $n-th$ market evolution ζ_n. To sum up, $\hat{\zeta} = (\hat{\zeta}_{i_s})^{i=1}_{s\in[0,1]}$ is a Walrasian variation economy which, as a simple calculation shows, ϵ–approximates the given evolution ζ, and whose equilibrium set is contained in $\Delta_\delta^{n-1} \times [0, 1]$.

Now, we are still left with the crucial question: can we show that $\hat{\zeta}$ possesses a well passable equilibrium path? Fortunately, as we have indicated above, all preparatory work needed for this has already been done in the previous Section. Actually, our situation here turns out to be a special case of the Situation in Section 11.1.1. Roughly speaking, this is due to the fact that the equilibrium set of $\hat{\zeta}$ is contained in $\Delta_\delta^{n-1} \times [0, 1]$ and $\hat{\zeta}$ is linear over any subspace, or say subprism, $\sigma \times [\rho_j, \rho_{j+1}]$ where σ is a simplex from the triangulation $\sum$ of Δ_δ^{n-1} and $[\rho_j, \rho_{j+1}]$ is a subinterval from the refined subdivision

R of $[0,1]$. To be more precise, consider $\hat{\zeta}|_{\sigma \times [\rho_j, \rho_{j+1}]} = \left(\hat{\zeta}_i |_{\sigma \times [\rho_j, \rho_{j+1}]} \right)_{i=1}^{n}$ for any $\sigma \times [\rho_j, \rho_{j+1}]$. The first $n-1$ component functions are linear in $p \in \sigma$ and in $s \in [\rho_j, \rho_{j+1}]$, and the equilibrium set of $\hat{\zeta}|_{\sigma \times [\rho_j, \rho_{j+1}]}$ is contained in $\cap_{i=1}^{n-1} \left(\hat{\zeta}_i |_{\sigma \times [\rho_j, \rho_{j+1}]} \right)^{-1} (0)$.

Thus, from the analysis from the last part of Section 11.1.1 we know that any connected component of the equilibrium set of $\hat{\zeta}|_{\sigma \times [\rho_j, \rho_{j+1}]}$ actually is finitely piecewise analytically path connected. Since the whole inner prism $\Delta_\delta^{n-1} \times [0,1]$ is decomposed into only finitely many subprisms $\sigma \times [\rho_j, \rho_{j+1}]$ which are, moreover, closed and therefore overlapping, this means that any connected equilibrium component of the whole evolution $\hat{\zeta}$ is finitely piecewise analytically path connected. Thus, particularly any joining equilibrium component is finitely piecewise analytically path connected, and we are done.

Perhaps the reader may feel a bit unsatisfied with our last reasoning having the impression that we took a sledgehammer to crack a nut. Actually, it seems to be immediately clear from geometrical intuition that the highly regular evolution $\hat{\zeta}$ cannot produce any unpassable, or in any other way pathological, equilibrium components. In fact, a closer look at the geometrical performance of the evolution $\hat{\zeta}$ confirms intuition to be right.

To this end again pick up any simplex $\sigma = <v_1, \ldots, v_n>$ from the triangulation $\sum$ of Δ_δ^{n-1} and any interval $[\rho_j, \rho_{j+1}]$ from the refined subdivision R of $[0,1]$. Now imagine geometrically the two image simplices $\overline{\sigma}_{1_{\rho_j}} := \hat{\zeta}_{1_{\rho_j}}(\sigma) \subset \sigma \times \mathbb{R} \subset \mathbb{R}^n$ and $\overline{\sigma}_{1_{\rho_{j+1}}} := \hat{\zeta}_{1_{\rho_{j+1}}}(\sigma) \subset \sigma \times \mathbb{R} \subset \mathbb{R}^n$ of σ under the first component function of $\hat{\zeta}$ at the two parametrizing points ρ_j and ρ_{j+1}. Recall that $\overline{\sigma}_{1_{\rho_j}}$ is the convex hull of the n image points $\zeta_{1_{\rho_j}}(v_n), \ldots, \zeta_{1_{\rho_j}}(v_n)$, and $\overline{\sigma}_{1_{\rho_{j+1}}}$ is the convex hull of the n image points $\zeta_{1_{\rho_{j+1}}}(v_1), \ldots, \zeta_{1_{\rho_{j+1}}}(v_n)$. By construction of the one–parametrization $((\hat{\zeta}|_\sigma)_s)_{s \in [\rho_j, \rho_{j+1}]}$ any vertex $\zeta_{1_{\rho_j}}(v_h)$ of $\overline{\sigma}_{1_{\rho_j}}, h = 1, \ldots, n$, moves linearly along the straight line $\{v_h\} \times \mathbb{R} \subset \sigma \times \mathbb{R} \subset \mathbb{R}^n$, which is orthogonal on σ in v_h, to the corresponding vertex $\hat{\zeta}_{1_{\rho_{j+1}}}(v_h)$ of the simplex $\overline{\sigma}_{1_{\rho_{j+1}}}$ for s running from ρ_j to ρ_{j+1}.

Now let us focus on the intersections of the moving image simplices $\overline{\sigma}_{1_s}, s \in [\rho_j, \rho_{j+1}]$, with the fixed domain simplex $\sigma \subset \Delta_\delta^{n-1}$. It is immediately clear from the construction that the union of all intersections recorded in the subprism $\sigma \times [\rho_j, \rho_{j+1}]$, i.e. $z_1 := \bigcup_{s \in [\rho_j, \rho_{j+1}]} (\overline{\sigma}_{1_s} \cap \sigma, s) \subseteq \sigma \times [\rho_j, \rho_{j+1}]$ consists of at most finitely many most well–behaved components which particularly are well passable.

Now take the image simplex $\overline{\sigma}_{2_{\rho_j}} := \hat{\zeta}_{2_{\rho_j}}(\sigma)$ under the second component function of $\hat{\zeta}$ and from the reiterated intersection set

$$z_2 := \bigcup_{s \in [\rho_j, \rho_{j+1}]} (\overline{\sigma}_{2_s} \cap z_1, s) \subseteq \sigma \times [\rho_j, \rho_{j+1}].$$

By construction the finitely many components of z_2 are equally well–behaved like those of z_1. Iterating this intersection process we finally arrive at

$$z_{n-1} := \bigcup_{s \in [\rho_j, \rho_{j+1}]} (\overline{\sigma}_{n-1_s} \cap z_{n-2}, s) \subseteq \sigma \times [\rho_j, \rho_{j+1}].$$

Due to Walras' law the equilibrium set of $\hat{\zeta}|_{\sigma \times [\rho_j, \rho_{j+1}]}$ equals z_{n-1}.

   To sum up we have convinced ourselves that under each successive step generating the equilibrium set z_{n-1} of $\hat{\zeta}|_{\sigma \times [\rho_j, \rho_{j+1}]}$ the well–behavedness of the connected components remains unaffected. Since there are only finitely many simplices σ and subintervals $[\rho_j, \rho_{j+1}]$ our findings show that *any component of the whole equilibrium set of $\hat{\zeta}$ is well–passable*.

11.2 Approximating Evolutions With Equilibrium Paths for the Other Basic Models From Part I

In the preceding Section we have presented two constructive methods how to approximate an evolution based on the Walrasian exchange model by Dierker by an evolution from this class which has well-behaved joining equilibrium components. Having reached this point it is dearly of natural interest how the presented methods can be adapted to evolutions based on the other models from Part I. Let us take these points one at a time.

(1) Both presented approximation methods immediately carry over to the *basic model of an exchange economy by Arrow and Hahn* (Section 1.1). Note particularly that the role of Δ_δ^{n-1} and $\Delta_{\delta/2}^{n-1}$ is now played by (n-1)-dimensional simplicial complexes D_δ^{n-1} and $D_{\delta/2}^{n-1}$ with

$$(\Delta^{n-1}\backslash U_L) \subset D_\delta^{n-1} \subset D_{\delta/2}^{n-1} \subset \Delta^{n-1}\backslash L$$

where $\mathcal{U}_L$ plays here the role which was played by $\mathring{\Delta}^{n-1}\backslash\Delta_\delta^{n-1}$ in our preceding constructions. More precisely, $\mathcal{U}_L$ is a relatively open neighborhood of the exception set L in Δ^{n-1} such that if $(p, s) \in (\mathcal{U}_L \cap \Delta_i^{n-1}) \times [0, 1]$ for some $i \in \{1, \ldots, n\}$ then $\zeta_{i_s}(p) > 2$. D_δ^{n-1} is a simplicial complex contained in $\Delta^{n-1}\backslash L$ with distance to L equal to or smaller than δ, and $D_{\delta/2}^{n-1}$ is an extension of D_δ^{n-1} with distance to L equal to or smaller than $\delta/2$. Figure 11.5 below gives an illustration of these notations.

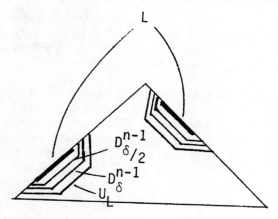

Fig. 11.5: Construction of the Domain of Approximating Evolutions with Equilibrium Paths

(2) Some more technicals efforts, however, are necessary to adapt our approximation methods from Section 11.1 to evolutions based on our basic model of a large exchange economy from Section 4.3.

In order to keep things easy we confine ourselves to a *compact* subspace $\widetilde{T}$ of the space $A^0 = D^0 \times (\mathbb{R}_+^n \backslash \{0^n\})$ of individual characteristics such that for some arbitrary, but fixed, $\epsilon > 0$ and some arbitrary, but fixed, triangulation Σ of Δ_ϵ^{n-1} any individual demand function $f : \mathring{\Delta}^{n-1} \times \mathbb{R}_{++} \to \mathbb{R}_+^n$ from $pr_{D^0}(\widetilde{T})$ is of the following form: for $i = 1, \ldots, n-1$ any restriction $f_i|_{\sigma \times]o,\alpha[}$ with $\sigma \in \Sigma$ and

$$\alpha := \max_{\substack{p \in \Delta_\epsilon^{n-1} \\ \omega \in pr_{\mathbb{R}_+^n \backslash \{0^n\}}(\widetilde{T})}} p\omega + 1$$

is a linear mapping. According to the budget equation $pf(p, p\omega) = p\omega$ one has $f_n(p, p\omega) = \frac{p\omega}{p_n} - \frac{1}{p_n} \sum_{i=1}^{n-1} p_i f_i(p, p\omega)$ for $\omega \in pr_{\mathbb{R}_+^n \backslash \{0^n\}}(\widetilde{T})$ and $p \in \Delta_\epsilon^{n-1}$.

On the neighborhood $\mathring{\Delta}^{n-1} \backslash \Delta_\epsilon^{n-1}$ of the boundary f has the usual properties and furthermore satisfies

$$\forall_{p \in \mathring{\Delta}^{n-1} \backslash \Delta_\epsilon^{n-1}} \forall_{\omega \in pr_{\mathbb{R}_+^n \backslash \{0^n\}}(\widetilde{T})} \exists_{i_{(p,\omega)} \in \{1,\ldots,n\}} f_{i_{(p,\omega)}}(p, p\omega) > \omega_{i_{(p,\omega)}}.$$

The latter condition just means that for any p near the boundary and for any initial endowment vector the demand for some $i - th$ good is greater than the initial endowment of good i. From this condition follows that any "individual" equilibrium price vector p_0 for any characteristics pair $(f, \omega) \in \widetilde{T}$, i.e. $f(p^0, p^0\omega) = \omega$, is in Δ_ϵ^{n-1}. Consequently, there is no equilibrium price vector in $\mathring{\Delta}^{n-1} \backslash \Delta_\epsilon^{n-1}$ for any large economy $\mu \in \mathcal{M}_{\widetilde{T}}$. Notice furthermore that any possible wealth $p^0\omega$ generated by any individual equilibrium price vector p^0 is in $]0, \alpha[$.

After these prerequisites let us see how to properly approximate a large evolution $(\mu_s)_{s \in [0,1]}$ with $\mu_s \in \mathcal{M}_{\widetilde{T}}$ for any $s \in [0, 1]$ such that nice joining equilibrium price paths obtain. Fortunately, this is straightforward. Since the large evolution $(\mu_s)_{s \in [0,1]}$ is a continuous path in the compact, complete, separable metric space $\mathcal{M}_{\widetilde{T}}$, it is natural to approximate it "polynomially" by finitely many convex large connection economies, i.e. by

$$M = ((1 - \tau)\mu_{s_j} + \tau\mu_{s_{j+1}})_{\substack{j=0,\ldots,k-1 \\ \tau \in [0,1]}}$$

where $s_0 = 0 < s_1 < \ldots < s_{k-1} < s_k = 1$. That M actually is a large economy evolution follows from our considerations in Section 12.2.

Now we still have to ensure that the piecewise linear approximating evolution M actually has nice joining equilibrium price paths. Nevertheless, this follows immediately from the following considerations. For any large s-state exchange economy $\mu_s, s \in [0, 1]$, the associate s-state excess demand function

$$\zeta_{\mu_s} : \mathring{\Delta}^{n-1} \longrightarrow \mathbb{R}^n$$

$$p \mapsto \int \underset{\sim}{f}(-, -, p)d\mu - \int \underset{\sim}{\omega}(-, -, p)d\mu$$

(with, as the reader will remember, $\underset{\sim}{f}(f\omega, p) = f(p, p\omega)$, and $\underset{\sim}{\omega}(f, \omega, p) = \omega$) is piecewise linear on Δ_ϵ^{n-1}, i.e. is linear on any $\sigma \in \Sigma$. This is easy to see. Firstly, any $\underset{\sim i}{f}, i = 1, \ldots, n - 1$, is linear on σ since for any $p^1, p^2, \alpha p^1 + \beta p^2 \in \sigma, \alpha, \beta \in \mathbb{R}$, one has $\underset{\sim i}{f}(f, \omega, \alpha p^1 + \beta p^2) = f_i(\alpha p^1 + \beta p^2, \omega(\alpha p^1 + \beta p^2)) = \alpha f_i(p^1, \omega p^1) + \beta f_i(p^2, \omega p^2)$ due to the properties of f. Secondly, $\int \underset{\sim}{f}(-, -, p)d\mu$ is linear on σ according to the following standard property of integrals: $\int h(x, \alpha p^1 + \beta p^2)dx = \alpha \int h(x, p^1)dx + \beta \int h(x, p^2)dx$. $\int \underset{\sim}{\omega}(-, -, p)d\mu$ trivially is linear on σ. Hence, to sum up, $\zeta_{\mu_s}(\alpha p^1 + \beta p^2) = \alpha\zeta_{\mu_s}(p^1) + \beta\zeta_{\mu_s}(p^2)$.

Finally, we again have to convince ourselves that also the one-parametrization of excess demand functions $\zeta_M = (\zeta_{\mu_s})_{s \in [0,1]}$ induced by the approximating large economy evolution M is composed of k convex transitions $(1 - \tau)\zeta_{\mu_{s_j}} + \tau\zeta_{\mu_{s_{j+1}}}, 0 \leq \tau \leq 1, j \in \{0, 1, \ldots, k - 1\}$. Formally,

$$\zeta_M = (\zeta_{\mu_s})_{s \in [0,1]} : \mathring{\Delta}^{n-1} \times [0, 1] \longrightarrow \mathbb{R}^n$$

$$(p, s) = (p, (1 - \tau)s_j + \tau s_{j+1}) \mapsto \zeta_{[(1-\tau)\mu_{s_j} + \tau\mu_{s_{j+1}}]}(p).$$

Now, by definition and from linearity of the integral operator for $\alpha, \beta \geq 0, \alpha + \beta = 1$ one has

$$
\begin{aligned}
\zeta_{(\alpha\mu_s + \beta\mu_{s'})}(p) &= \int \underset{\sim}{f}(-,-,p) d(\alpha\mu_s + \beta\mu_{s'}) \\
&\quad - \int \underset{\sim}{\omega}(-,-,p) d(\alpha\mu_s + \beta\mu_{s'}) \\
&= \alpha \int \underset{\sim}{f}(-,-,p) d\mu_s + \beta \int \underset{\sim}{f}(-,-,p) d\mu_{s'} \\
&\quad - \alpha \int \underset{\sim}{f} d\mu_s - \beta \int \underset{\sim}{\omega} d\mu_{s'} \\
&= \alpha\zeta_{\mu_s} + \beta\zeta_{\mu_{s'}}.
\end{aligned}
$$

Particularly, this means

$$
\zeta_{((1-\tau)\mu_{s_j} + \tau\mu_{s_{j+1}})} = (1-\tau)\zeta_{\mu_{s_j}} + \tau\zeta_{\mu_{s_{j+1}}}
$$

for any $\tau \in [0,1]$ and $j \in \{0, 1, \ldots, k-1\}$. With each $\zeta_{\mu_{s_j}}, j = 0, \ldots, k-1$ being linear on any $\sigma \in \Sigma$, we are done.

(3) There is no difficulty at all to carry over our previous analysis for Dierker's version of a Walrasian exchange model to the two models from the framework without Walras' law and homogeneity from Chapter 5. The only difference is that now the approximation area is $T_{\gamma/2}^{n,\alpha+1}$, and not $\Delta_{\delta/2}^{n-1}$.

(4) Since the analysis of the two basic models from the equilibrium framework with production, taxes, and subsidies from Chapter 5 is considerably more complicated than that of the other basic models we defer it to the end of this section. Thus let us look now at the two basic models from the temporary equilibrium framework from Chapter 7. To start with the quantity constrained micromodel with effective demand à la Benassy (Section 7.1) there is no difficulty to adapt both of our approximation methods. As to the piecewise linear approximation method one just has to be careful with choosing a suitable triangulation Σ_s of $(G_-)^m \times (G_+)^m$ for approximating an s-state economy $(p_s; \widetilde{z}_s^1, \ldots, \widetilde{z}_s^m; F_s^1, \ldots, F_s^m)$. (The restriction of Σ_s to $G_- \times G_+ \subset \partial[(G_-)^m \times (G_+)^m]$ is the relevant triangulation for the function $\widetilde{z}_s$.) For, due to the construction of the equilibrium equivalent self-mapping g a suitable triangulation Σ_s must obviously count for the respective zero sets of the $3nm$ functions $\max(\underline{z}_i^a; \widetilde{z}_i(p_i; \underline{z}^a; \bar{z}^a))$, $\min(\bar{z}_i^a; \widetilde{z}_i^a(p_i; \underline{z}^a; \bar{z}^a))$, and $\min(0; \widetilde{z}_i^a(p; \underline{z}^a, \bar{z}^a))$. More precisely, having chosen some triangulation of $(G_-)^m \times (G_+)^m$ which admits an ϵ-approximation of the functions $\widetilde{z}_s^1, \ldots, \widetilde{z}_s^m; F_s^1, \ldots, F_s^m$ one still has to take into account the special functions forming the equilibrium self-mapping $g : (G_-)^m \times (G_+)^m \longrightarrow (G_-)^m \times (G_+)^m$. (Recall that the zero set of a linear function restricted to a k-dimensional simplex is a simplex of dimension $\leq k$.) But, obviously,

this is straightforward. Replacing g by $g - id$ all previous considerations also apply to this basic model.

(5) Now we come to the multi-sectoral quantity constrained basic model from Section 7.2. As we will see, the adaptation of the piecewise linear approximation method is straightforward. For any s-state economy

$$E_s := (l_{1_s}^s(-), l_{1_s}^d(-), y_{1_s}^s(-), y_{1_s}^d(-); \ldots;$$
$$l_{m_s}^s(-), l_{m_s}^d(-), y_{m_s}^s(-), y_{m_s}^d(-))$$

from a given evolution a piecewise linear approximation is achieved in the following way. Choose any sector i. Each of the 4 sectoral reaction functions $l_{i_s}^{s,d}(l_1, \ldots, l_{i-1}, l_{i+1}, \ldots l_m; y_1, \ldots, y_i, \ldots, y_m)$ and $y_{i_s}^{s,d}(l_1, \ldots, l_i, \ldots, l_m; y_1, \ldots, y_{i-1}, y_{i+1}, \ldots, y_m)$ is a continuous function of $2m - 1$ variables. More precisely, the two functions $l_{i_s}^{s,d}(l_1, \ldots, l_{i-1}, l_{i+1}, \ldots, l_m; y_1, \ldots, y_i, \ldots, y_m)$ are defined on

$$[0, y_{i_{max}}] \times C_i := [0, y_{i_{max}}] \times [0, l_{1_{max}}] \times \ldots \times [0, l_{i-1_{max}}]$$
$$\times [0, l_{i+1_{max}}] \times \ldots \times [0, l_{m_{max}}]$$
$$\times [0, y_{1_{max}}] \times \ldots \times [0, y_{i-1_{max}}]$$
$$\times [0, y_{i+1_{max}}] \times \ldots \times [0, y_{m_{max}}].$$

And correspondingly, the two functions $y_{i_s}^{s,d}(l_1, \ldots, l_i, \ldots, l_m; y_1, \ldots, y_{i-1}, y_{i+1}, \ldots, y_m)$ are defined on $[0, l_{i_{max}}] \times C_i = [0, l_{i_{max}}] \times [0, l_{1_{max}}] \times \ldots \times [0, l_{i-1_{max}}] \times [0, l_{i+1_{max}}] \times \ldots \times [0, l_{m_{max}}] \times [0, y_{1_{max}}] \times \ldots \times [0, y_{i-1_{max}}] \times [0, y_{i+1_{max}}] \times \ldots \times [0, y_{m_{max}}]$. Now choose some finite triangulations $\sum_s^1$ and $\sum_s^2$ of the domains $[0, y_{i_{max}}] \times C_i$ and $[0, l_{i_{max}}] \times C_i$ respectively such that the 4 finitely piecewise linear functions $\bar{l}_{i_s}^{s,d}(-)$ and $\bar{y}_{i_s}^{s,d}(-)$ which are formed in complete analogy to our previous constructions, $\epsilon/2$-approximate the four given functions $l_{i_s}^{s,d}$ and $y_{i_s}^{s,d}(-)$ on their respective domains. Unfortunately, there may still arise a difficulty. There may be argument tuples $(l_1', \ldots, l_{i-1}', l_{i+1}', \ldots, l_m'; y_1', \ldots, y_{i-1}', y_{i+1}', \ldots, y_m')$ from C_i for which the associate state diagram in the $l_i - y_i$–coordinate box formed by the 4 reaction functions parametrized with this argument tuple, i.e.

$$l_{i_s}^s(l_1', \ldots, l_{i-1}', l_{i+1}', \ldots, l_m'; y_1', \ldots, y_{i-1}', y_i, y_{i+1}', \ldots, y_m'),$$
$$l_{i_s}^d(l_1', \ldots, l_{i-1}', l_{i+1}', \ldots, l_m'; y_1', \ldots, y_{i-1}', y_i, y_{i+1}', \ldots, y_m'),$$
$$y_{i_s}^s(l_1', \ldots, l_{i-1}', l_i, l_{i+1}', \ldots, l_m'; y_1', \ldots, y_{i-1}', y_{i+1}', \ldots, y_m'),$$
$$y_{i_s}^d(l_1', \ldots, l_{i-1}', l_i, l_{i+1}', \ldots, l_m'; y_1', \ldots, y_{i-1}', y_{i+1}', \ldots, y_m'),$$

is not of the required form, i.e. does not consist of two wedges constituting a *unique* sectoral equilibrium. But this can easily be removed by appropriately modifying the $\epsilon/2$-approximating piecewise linear functions $\bar{l}_{i_s}^{s,d}(-)$ and $\bar{y}_{i_s}^{s,d}(-)$ such that $\epsilon/2$-approximating finitely piecewise linear functions $\bar{\bar{l}}_{i_s}^{s,d}(-)$ and $\bar{\bar{y}}_{i_s}^{s,d}(-)$ obtain which also satisfy this requirement. This is always possible since the domain triangulations $\sum_s^1$ and $\sum_s^2$ are finite. Let us denote the obtained $\epsilon/2$-approximating economy by $\bar{\bar{e}}_s$.

Clearly, the same difficulty may arise when we perform the convex transition between any two $\epsilon/2$-approximating economies $\bar{\bar{E}}_s$ and $\bar{\bar{E}}_{s'}$. Again, we can overcome this difficulty thanks to the finiteness by slightly modifying the finitely piecewise linear reaction functions of $\bar{\bar{E}}_s$ and $\bar{\bar{E}}_{s'}$ and putting in finitely many additional intermediate approximating economies.

Now, everything works as before.

(6) We are now left with the two basic models from the general equilibrium framework with production, taxes, and subsidies from Chapter 6. While for the other basic models geometrically elementary modifications were sufficient to yield an adaptation, this time we must resort to further results from algebraic geometry and algebraic topology.

The following Theorem gives the desired results.

Theorem 11.8.
For any evolution $(\zeta_s, t_s, A_s, A_s^*)_{s\in[0,1]}$ basing on the model with production and tax schemes from Section 6.1 and any evolution $(\zeta_s, t_s, A_s, A_s^*, A_s^{**})_{s\in[0,1]}$ basing on the model with production and tax and subsidy schemes from Section 6.2, and for any $\epsilon > 0$ there exist ϵ-uniformly approximating evolutions

$$(\hat{\zeta}_s, \hat{t}_s, \hat{A}_s, \hat{A}_s^*)_{s\in[0,1]} \ and$$
$$(\hat{\zeta}_s, \hat{t}_s, \hat{A}_s, \hat{A}_s^*, \hat{A}_s^{**})_{s\in[0,1]}$$

which have nice, i.e. finitely piecewise analytical, equilibrium paths. Moreover, if A_s (A_s^*, A_s^{**}) is a finitely piecewise analytical path in $\mathbb{R}^{nm}$, then one can even choose $\hat{A}_s = A_s$ $(\hat{A}_s^* = A_s^*, \hat{A}_s^{**} = A_s^{**})$.

The proof of Theorem 11.8 is relegated to Appendix A at the end of the monograph. In a nutshell the proof provides an appropriate adaptation of the polynomial approximation method to our present situation.

11.3 A Strong Connectedness Result for the Graphs of the Equilibrium Correspondences of the Basic Models From Part I

In this Section we will show that our conception of an evolution of economies immediately leads to a certain strong global connectedness result for the graph of the equilibrium correspondence of each basic model from Part I. Particularly, our findings go far beyond the well-known results about the manifold property of the graph of the Walras correspondence developed by Balasko and others in the seventies and eighties (Balasko 1988).

What we are aiming for in this Section is to present an intuitive and unifying construction based on connection evolutions which shows for every of our basic set-ups that any two pairs (E_0, P^0) and (E_1, P^1) of an associated economy and an equilibrium, or say any two points from the graph of the equilibrium correspondence, can be connected in a simple way by such pairs, that means *within* the graph of the equilibrium correspondence.

Actually, our method not only allows to considerably strengthen the well-known global connectedness results on the structure of the graph of the Walras correspondence by Balasko and others. It is also simple and economically appealing. We will come back to the precise relationship of our findings to the investigations on the graph of the Walras correspondence in the literature in Section 14.1 below.

Generally, our method is based on the following principles. We first construct well-understood auxiliary (intermediate) economies E_0' and E_1' containing P^0 and P^1 in their respective equilibrium sets. A particularly simple example are economies E_0', E_1' with *unique* equilibria P^0, P^1 respectively. Clearly, the canonical connection evolutions from E_0 to E_0' provided in Section 12.1 possess a 'stationary' equilibrium path in P^0. The analogous consideration applies to E_1 and E_1'. Now it just remains to connect E_0' with E_1' by a well-understood connection evolution. As we will see, for some of our basic models this is straightforward, while for the others it becomes a piece of handcraft. However, in any case the result is intuitive from both the economic and the geometrical viewpoint. We will proceed in 7 steps dealing with our various basic set-ups from Part I one after the other.

(1) Let us this time first explicitly develop the whole method for our basic models without Walras' law and homogeneity. Afterwards we will point to the modifications which are necessary to adapt it to the other models.

Let any two pairs (ζ_0, p^0) and (ζ_1, p^1) of an exchange economy without Walras' law and homogeneity of type I and an associate equilibrium price vector from $\mathbb{R}^n_{++}$ be given. To construct an auxiliary economy $\zeta'_0 = (\zeta'_{i_0})^n_{i=1}$ with unique equilibrium p^0 consider the n affine (n-1)-dimensional coordinate hyperplanes $H_1(p^0), \ldots, H_n(p^0)$ of $\mathbb{R}^n \times \mathbb{R}$ containing the point p^0. Of course, $\{p^0\}$ is precisely their intersection set. Now choose n functions $h_{i_0} : \mathbb{R}^n \longrightarrow \mathbb{R}$, $i = 1, \ldots, n$, with the following properties: the graph of h_{i_0} is an n-dimensional hyperplane of $\mathbb{R}^n \times \mathbb{R}$ with zero set equal to $H_i(p^0) \cap \mathbb{R}^n$, $h_{i_0}(x) > 0$ for any x with $x_i < p^0_i$, and $h_{i_0}(x) < 0$ for any x with $x_i > p^0_i$. The candidates for our component functions ζ'_{i_0}, $i = 1, \ldots, n$, are the restrictions $h_{i_0}|_{\mathbb{R}^n_+}$. However, we still have to bound the functions $h_{i_0}|_{\mathbb{R}^n_+}$ from below by some arbitrary $b \in \mathbb{R}^n_{--}$. Thus, $\zeta'_{i_0}(x) = \max(h_{i_0}(x), b_i)$ for $x \in \mathbb{R}^n_+$. Evidently, ζ'_0 is a basic economy with unique equilibrium p^0.

Now let us construct an auxiliary basic economy ζ'_1 for ζ_1 in complete analogy. Our next aim is to provide a well-understood connection evolution between ζ'_0 and ζ'_1 which particularly possesses a well-understood equilibrium path connecting p^0 and p^1. To this end let us consider the segment $((1-s)p^0 + sp^1, s)$ in $\mathbb{R}^n_+ \times [0,1]$ from $(p^0, 0)$ to $(p^1, 1)$. Now we are in the position to propose the following candidate for a connection evolution between $\bar{\zeta}'_0$ and $\bar{\zeta}_1$: any s-state economy is constructed on the same principle as for ζ'_0 and ζ'_1 with the main difference that now $(1-s)p^0 + sp^1$ is the unique equilibrium. Clearly, this yields a connection evolution.

Composing this connection evolution with the convex connection evolutions from ζ_0 to ζ'_0 and from ζ_1 to ζ'_1, respectively, obviously achieves a twofold composite connection evolution from ζ_0 to ζ_1 with a finitely piecewise linear equilibrium path consisting of the three segments (p^0, s) for $0 \leq s \leq 1/3$, $((1-3(s-1/3))p^0 + 3(s-1/3)p^1, s)$ for $1/3 \leq s \leq 2/3$, and (p^1, s) for $2/3 \leq s \leq 1$. This completes our construction. Figure 11.6 gives an illustration.

(2) With easy modifications the described method can immediately be adapted to the basic Walrasian exchange models by Dierker and by Arrow/Hahn.

(3) For the basic model of a large exchange economy an adaptation is equally straightforward. Given two pairs (μ_0, p^0) and (μ_1, p^1) one chooses for $\mu'_0(\mu'_1)$ that probability distribution on T which concentrates its whole mass on a pair (f^0, ω) $((f^1, \omega^1))$ with an excess demand function $f^0(-) - \omega^0(f^1(-) - \omega^1)$ of an analogous form to $\zeta'_0(\zeta'_1)$ above. Thus p^0 (p^1) is the unique equilibrium price vector for the large economy μ'_0 (μ'_1). With the connection evolution between μ'_0 and μ'_1 whose s-state economies give the whole mass to analogously shifted pairs (f^s, ω^s) as before, $0 < s < 1$, we are done. (To be more specific, the derived s-state excess demand function $f^s(-) - \omega^s$ must have the unique equilibrium vector $(1-s)p^0 + sp^1$).

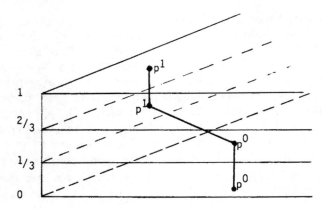

Fig. 11.6: Construction Scheme of a Connecting Economy for the Basic Model without Homogeneity and Walras' Law

(4) Now we come to the general equilibrium framework with production, taxes, and subsidies from Chapter 6. Let us begin with the first basic model with production and taxes from Section 6.1. Let two pairs $((\zeta_0, t_0, A_0, A_0^*), (p_0, r_0))$ and $((\zeta_1, t_1, A_1, A_1^*), (p_1, r_1))$ of an economy and an associate equilibrium be given. We will provide an intuitive general construction of a 4-fold convex composite connection evolution whose equilibrium set particularly contains the "pennant" of the following Figure 11.7. To sim-

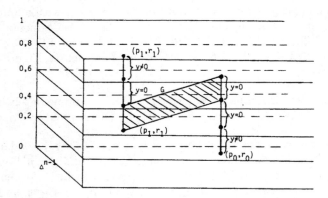

Fig. 11.7: Construction Scheme of a Connecting Economy and Equilibrium Subset for the Basic Model with Production, Taxes, and Subsidies I

plify notations we will consider excess demand functions and tax functions as defined on $\Delta^{n-1} \times \mathbb{R}_+$ (rather than on $(\mathbb{R}_+^n \setminus \{0\}) \times \mathbb{R}_+$). We further denote the segment $(1 - \lambda)(p_0, r_0) + \lambda(p^1, r^1)$, $0 \le \lambda \le 1$ from (p_0, r_0) to (p_1, r_1) in $\Delta^{n-1} \times \mathbb{R}_+$ by G.

Now we are going to provide two auxiliary economies for each of the two given economies such that the equilibrium set of the first auxiliary economy contains (p_0, r_0) $((p_1, r_1))$, and the equilibrium set of the second one contains the segment G. Moreover, the 4-fold convex transition yields a connection evolution.

Actually there are several ways to achieve that. Here we propose the following one which has the virtue of being both simple and economically appealing. Let us denote the two addressed auxiliary economies for the first of the two given economies $(\zeta_0, t_0, A_0, A_0^*)$ by $(\zeta_{0.2}, t_{0.2}, A_{0.2}, A_{0.2}^*)$ and $(\zeta_{0.4}, t_{0.4}, A_{0.4}, A_{0.4}^*)$, and those for $(\zeta_1, t_1, A_1, A_1^*)$ by $(\zeta_{0.8}, t_{0.8}, A_{0.8}, A_{0.8}^*)$ and $(\zeta_{0.6}, t_{0.6}, A_{0.6}, A_{0.6}^*)$, respectively.

Let us choose two arbitrarily small and bounded relative neighborhoods $U_1 \subsetneq U_2$ of G in $\Delta^{n-1} \times \mathbb{R}_+$. Now we can continuously glue the zero function on $\Delta^{n-1} \times \mathbb{R}_+$ with some arbitrary excess demand function $\widetilde{\zeta}$ such that the obtained function equals the zero function on U_1 and $\widetilde{\zeta}$ on $U_2^c = (\Delta^{n-1} \times \mathbb{R}_+) \backslash U_2$. The obtained function will serve us as excess demand function $\zeta_{0.2} = \zeta_{0.4} = \zeta_{0.6} = \zeta_{0.8}$. Moreover, $t_{0.2} = t_{0.4} = t_{0.6} = t_{0.8}$ with $t_{0.2}(p, r) := r - p\zeta_{0.2}(p, r)$ clearly has an admissible shape (particularly, it is nowhere negative on its domain).

With the production and after-tax matrices, however, our choice has to be more sophisticated. For the 0.2- and 0.8-auxiliary economy it is easy: we just take $A_{0.2} = A_0$, $A_{0.2}^* = A_{0.4}$, and $A_{0.8} = A_1$, and $A_{0.8}^* = A_1^*$. Likewise, we take A_0 for $A_{0.4}$, and A_1 for $A_{0.6}$. As after-tax matrix $A_{0.4}^*$ choose some arbitrary matrix which fulfills

(i) $p' A_{0.4}^* \le 0$ for any $(p, r) \in G'$, and

(ii) $A_{0.4}^* \le A_{0.4} = A_0$ such that any entry $a_{0,4_{ij}}^*$ is generated in the following way: there is a $\tau_{ij} \in [0, 1]$ with

$$a_{0,4_{ij}}^* = a_{0_{ij}} - \tau_{ij}|a_{0_{ij}}|,$$

where $A_0 = (a_{0_{ij}})_{ij}$ and $A_{0.4}^* = (a_{0,4_{ij}}^*)_{ij}$.

Accordingly choose some suitable matrix. (Evidently there is no difficulty with all these choices.)

From Section 12.1 we know that the 4-fold convex transition yields a connection evolution from $(\zeta_0, t_0, A_0, A_0^*)$ to $(\zeta_1, t_1, A_1, A_1^*)$. (The space of economies is convex.)

Moreover, the following brief consideration shows us that the equilibrium set of this 4-fold composite connection evolution really contains the "pennant" from Figure 11.7 . Let us begin with the first convex connection evolution from 0 to 0.2. Its equilibrium set contains the segment $(p_0, r_0) \times [0; 0.2]$ since by construction for $0 \leq \sigma \leq 1$

$$(1 - \sigma)\zeta_0(p_0, r_0) + \sigma\zeta_{0.2}(p_0, r_0) \stackrel{(E.2)}{=} (1 - \sigma)A_0 y_0$$

$$= A_0[(1 - \sigma)y_0] = [(1 - \sigma)A_0 + \sigma A_{0.2}][(1 - \sigma)y_0]$$

for some suitable $y_0 \in \mathbb{R}_+^m$.

The reader will have noticed that for the sake of simpler notation we consider the connection evolution from 0 to 0.2 for the moment as quasi standing by itself, that means with an evolution parameter s running from 0 to 1 instead of running from 0 to 0.2. We will adhere to this simplification in the following.

For the second convex connection evolution from 0.2 to 0.4 we have

$$(1 - \sigma)\zeta_{0.2}(\pi_0, r_0) + \sigma\zeta_{0.4}(\pi_0, r_0) = (1 - \sigma)0 + \sigma 0$$

$$= A_{0.4}0 = A_0 0 = [(1 - \sigma)A_{0.2} + \sigma A_{0.4}]0 = 0 \in \mathbb{R}^m.$$

The cases 0.6 to 0.8 and 0.8 to 1 are completely analogous. For the middle convex connection from 0.4 to 0.6 we have for any $(p, r) \in G$ the relations

$$(1 - \sigma)\zeta_{0.4}(p, r) + \sigma\zeta_{0.6}(p, r) = (1 - \sigma)0 + \sigma 0$$

$$= [(1 - \sigma)A_{0.4} + \sigma A_{0.6}]0 = 0 \in \mathbb{R}^m.$$

Note particularly that due to our constructions there are no problems with the equilibrium conditions E.1 and E.3.

This completes our proof that the well-behaved pennant from Figure 11.7 is contained in the equilibrium set of our 4-fold composite convex connection evolution.

(5) Now let us tackle the second basic model with production, taxes, and subsidies from Section 6.2. Remember that now the domain of the economic state functions is $\Delta^{n-1} \times \mathbb{R}_+ \times [0, 1]$, and equilibria are confined to $D_1 \cup D_2 := \{(p, 0, \gamma)\} \cup \{(p, r, 1)\} \subset \Delta^{n-1} \times \mathbb{R}_+ \times [0, 1]$.

Let two pairs $((\zeta_0, t_0, A_0, A_0^*, A_0^{**}), (p_0, r_0, \gamma_0))$ and $((\zeta_1, t_1, A_1, A_1^*, A_1^{**}), (p_1, r_1, \gamma_1))$ of an economy and an associate equilibrium be given.

We will distinguish two main cases.

(1) both equilibria (p_0, r_0, γ_0) and (p_1, r_1, γ_1) lie in the closed space D_2, i.e. $\gamma_0 = \gamma_1 = 1$,

(2) they do not.

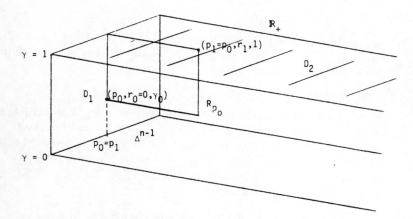

Fig. 11.8: Construction Scheme of a Connecting Economy and Equilibrium Subset for the Basic Model with Production, Taxes, and Subsidies II

In *case (1)* everything works in complete analogy to the preceding model. Just notice that $p((1-s)A_0^* + sA_1^* + (1-s)\gamma A_0^{**} + s\gamma A_1^{**}) = (1-s) \cdot p(A_0^* + A_1^{**})$ for any $(p, r, \gamma) \in G := \{(1-\tau)(p_0, r_0, \gamma_0) + \tau(p_1, r_1, \gamma_1) = (1-\tau)(p_0, r_0, 1) + \tau(p_1, r_1, 1), \tau \in [0, 1]\}$. One has just to be careful that the neighborhood U_2 of the segment G in the new domain $\Delta^{n-1} \times \mathbb{R}_+ \times [0, 1]$ does not intersect the face $\Delta^{n-1} \times \mathbb{R}_+ \times \{0\}$ where we have the strict positivity of the tax function. This is easily achieved, for instance, by requiring that $\gamma \geq 1/2$ for any $(p, r, \gamma) \in U_2$.

In *case (2)*, however, we now have also to take into account the additional requirement that $t(p, r, 0) = r - p\zeta(p, r, 0) > 0$ for all $(p, r) \in \Delta^{n-1} \times \mathbb{R}_+$. Let us briefly recall the equilibrium conditions for an equilibrium $(\hat{p}, \hat{r}, \hat{\gamma})$

(E'.1) $\hat{p}(A^* + \hat{\gamma}A^{**}) \leq 0$

(E'.2) $\zeta(\hat{\pi}, \hat{r}, \hat{\gamma}) = A\hat{y}$ for some $\hat{y} \in \mathbb{R}_+^m$

(E'.3) $\hat{p}'(A^* + \hat{\gamma}A^{**})\hat{y} = 0$.

Let us first consider the *special case* that $p_0 = p_1$, $r_0 = 0$, $0 \leq \gamma_0 < 1$, and $r_1 \geq 0$, $\gamma_1 = 1$. All other cases will turn out to be simple compositions of this and the preceding case. See for an illustration of the following constructions Figure 11.8 above.

Let us distinguish *two further subcases* $\gamma_0 = 0$ and $0 < \gamma_0 < 1$. More formally, $(p_1, r_1, \gamma_1) = (p_0, r_1, 1)$, and (i) $(p_0, r_0, \gamma_0) = (p_0, 0, 0)$, and (ii) $(p_0, r_0, \gamma_0) = (p_0, 0, \gamma_0)$.

(i) Let us begin with the case $\gamma_0 = 0$. We first construct an auxiliary economy $(\zeta_{0.5}, t_{0.5}, A_{0.5}, A_{0.5}^*, A_{0.5}^{**})$ whose equilibrium set contains the whole rectangular

$$R_{p_0} := \{(p_0, r, \gamma) | 0 \le r \le r_1, \gamma_0 \le \gamma \le 1\} = \{p_0\} \times [0, r_1] \times [\gamma_0, 1]$$

in $\Delta^{n-1} \times \mathbb{R}_+ \times [0, 1]$. (Notice that the two given equilibria $(p_0, r_0, \gamma_0) = (p_0, 0, 0)$ and $(p_1, r_1, \gamma_1) = (p_0, r_1, 1)$ are diagonally opposite corner points of the rectangular R_{p_0}.) To this end choose any admissible excess demand function $\zeta_{0.5}$ with the additional property that $\zeta_{0.5}(p_0, r, \gamma) = \zeta_0(p_0, r_0, \gamma_0)$ for $r_0 = 0 \le r \le r_1$ and $\gamma_0 = 0 \le \gamma \le 1$, i.e. on R_{p_0} (see Figure 11.8).

Actually, there is no difficulty with this choice. For instance, one can obtain an admissible function $\zeta_{0.5}$ by gluing the constant function $c(p, r, \gamma) = \zeta_0(p_0, 0, 0)$ on $\Delta^{n-1} \times \mathbb{R}_+ \times [0, 1]$ over some open neighborhood U of the rectangle R_{p_0} with ζ_0. The choice of $\zeta_{0.5}$ determines $t_{0.5}(p, r, \gamma) = r - p\zeta_{0.5}(p, r, \gamma)$. Furthermore, choose $A_{0.5} = A_0$, $A_{0.5}^* = A_0^*$, and $A_{0.5}^{**} = 0$. It is easy to cheek that $(p_0, r_0, \gamma_0) = (p_0, 0, 0)$ is contained in the equilibrium set of *any* state economy of the convex connection evolution from $(\zeta_0, t_0, A_0, A_0^*, A_0^{**})$ to

$$(\zeta_{0.5}, t_{0.5}, A_{0.5}, A_{0.5}^*, A_{0.5}^{**}) = (\zeta_{0.5}, t_{0.5}, A_0, A_0^*, 0).$$

Furthermore, it is evident that R_{p_0} is contained in the equilibrium set of $(\zeta_{0.5}, t_{0.5}, A_{0.5}, A_{0.5}^*, A_{0.5}^{**})$.

With the two pairs
$((\zeta_{0.5}, t_{0.5}, A_{0.5}, A_{0.5}^*, A_{0.5}^{**}), (p_1, r_1, \gamma_1) = (p_0, r_1, 1))$ and
$((\zeta_1, t_1, A_1, A_1^*, A_1^{**}), (p_1, r_1, \gamma_1))$ we are in the situation of case (1) above.

(ii) Let us now tackle the case $0 < \gamma_0 < 1$. The equilibrium conditions satisfied by $(p_0, 0, \gamma_0)$ take the following form:

(E'.1) $p_0'(A_0^* + \gamma_0 A^{**}) \le 0$

(E'.2) $\zeta_0(p_0, 0, \gamma_0) = A_0 \hat{y}_0$ for some $\hat{y}_0 \in \mathbb{R}_+^m$

(E'.3) $p_0'(A_0^* + \gamma_0 A_0^{**})\hat{y}_0 = 0$.

By definition $\gamma_0 A^{**} = (\gamma_0 \chi_{0_{ij}} |a_{0_{ij}}|)_{ij}$ with $\chi_{0_{ij}} \in [0, 1]$

As before we construct an auxiliary economy $(\zeta_{0.5}, t_{0.5}, A_{0.5}, A_{0.5}^*, A_{0.5}^{**})$. $\zeta_{0.5}$ is constructed in complete analogy to the preceding case. (Since now $\gamma_0 > 0$, one has also to take care of the requirement $t_{0.5}(p_0, 0, 0) = -p_0 \zeta_{0.5}(p_0, 0, 0) > 0$ when one employs the gluing method mentioned above. But this is easily ensured by choosing U in a way that its distance to $\Delta^{n-1} \times \mathbb{R}_+ \times \{0\}$ is positive.) Again, as before, $A_{0.5} = A_0$, $A^*_{0.5} = A_1$. However, this time we choose $A^{**}_{0.5} = \gamma_0 A^{**}_0$.

Now consider the following connection evolution from $(\zeta_0, t_0, A_0, A^*_0, A^{**}_0)$ to $(\zeta_{0.5}, t_{0.5}, A_{0.5}, A^*_{0.5}, A^{**}_{0.5}) = (\zeta_{0.5}, t_{0.5}, A_0, A^*_0, \gamma_0 A^{**}_0)$: for ζ, t, A, and A^* just choose the *convex transitions*. For the transition from $A^{**}_0 = (\chi_{0_{ij}} | a_{0_{ij}} |)_{ij}$ to $A^{**}_{0.5} = \gamma_0 A^{**}_0$ choose $\chi_{s_{ij}} := \frac{\gamma_0}{(1-2s)\gamma_0 + 2s} \chi_{0_{ij}}$ for any $s \in [0, 1/2]$. Notice that the choice has been made such that

$$\gamma_s A^{**}_s = \left([(1-2s)\gamma_0 + 2s] \frac{\gamma_0}{(1-2s)\gamma_0 + 2s} \chi_{0_{ij}} | a_{0_{ij}} | \right)_{ij}$$
$$= (\gamma_0 \chi_{0_{ij}} | a_{0_{ij}} |)_{ij} = \gamma_0 A^{**}_0$$

for any $s \in [0, 1/2]$.

It is immediate that $(p_0, r_0, \gamma_0) = (p_0, 0, \gamma_0)$ is contained in the equilibrium set of any s-state economy of this connection evolution. Analogously to before, the whole rectangle $R_{p_0} = \{(p_0, r, \gamma) | 0 \leq r \leq r_1, \gamma_0 \leq \gamma \leq 1\} = \{p_0\} \times [0, r_1] \times [\gamma_0, 1] \subset \Delta^{n-1} \times \mathbb{R}_+ \times [0, 1]$ is contained in the equilibrium set of $(\zeta_{0.5}, t_{0.5}, A^*_{0.5}, A^{**}_{0.5})$.

With the two pairs $((\zeta_{0.5}, t_{0.5}, A_{0.5}, A^*_{0.5}, A^{**}_{0.5}), (p_1, r_1, \gamma_1) = (p_0, r_1, 1))$ and $((\zeta_1, t_1, A_1, A^*_1, A^{**}_1), (p_1, r_1, \gamma_1))$ we are again in the situation of case (1).

So far, however, we have only considered the special case that $p_0 = p_1$, $r_0 = 0, 0 \leq \gamma_0 < 1$, and $r_1 \geq 0, \gamma_1 = 1$. Nevertheless, all other possible cases can easily be settled by simple compositions of the constructions provided so far (see Figure 11.9 below):

(a) $p_0 = p'_1, r_0 = r'_1 = 0, 0 \leq \gamma_0 < 1, 0 < \gamma'_1 < 1$ (trivial)

(b) $p_0 \neq p''_1, r_0 = r''_1 = 0, 0 \leq \gamma_0, \gamma''_1 < 1$.

Just construct for the two pairs $((\zeta_0, t_0, A_0, A^*_0, A^{**}_0), (p_0, 0, \gamma_0))$ and $((\zeta_1, t_1, A_1, A^*_1, a^{**}_1), (p_1, 0, \gamma_1))$ the two auxiliary canonical economies $(\zeta^I_{0.5}, t^I_{0.5}, A^I_{0.5}, A^{*I}_{0.5}, A^{**I}_{0.5})$ and $(\zeta^{II}_{0.5}, t^{II}_{0.5}, A^{II}_{0.5}, A^{*II}_{0.5}, A^{**II}_{0.5})$. Then $R_{p_0} = \{p_0\} \times \{0\} \times [\gamma_0, 1]$ and $R_{p_1} = \{p''_1\} \times \{0\} \times [\gamma''_1, 1]$. With the two pairs $((\zeta^I_{0.5}, t^I_{0.5}, A^I_{0.5}, A^{*I}_{0.5}, A^{**I}_{0.5}), (p_0, 0, 1))$

and $((\zeta_{0.5}^I, t_{0.5}^I, A_{0.5}^I, A_{0.5}^{*I}, A_{0.5}^{**I}), (p_1'', 0, 1))$ one again is in the situation of case (1).

(c) $p_0 \neq p_1''', r_0 = 0, r_1''' > 0, 0 \leq \gamma_0 < 1, \gamma_1''' = 1.$

Construct for $((\zeta_0, t_0, A_0, A_0^*, A_0^{**}), (p_0, 0, \gamma_0))$ the auxiliary economy $(\zeta_{0.5}, t_{0.5}, A_{0.5}, A_{0.5}^*, A_{0.5}^{**})$.

Then $R_{p_0} = \{p_0\} \times \{0\} \times [\gamma_0, 1]$. With the two pairs $((\zeta_{0.5}, t_{0.5}, A_{0.5}, A_{0.5}^*, A_{0.5}^{**}), (p_0, 0, 1))$ and $((\zeta_1, t_1, A_1, A_1^*, A_1^{**}), (p_1''', r_1''', 1))$ we are again in the situation of case (1).

The reader should note well, however, that the interval $[0, 1]$ in Figure 11.9 below represents the space of subsidy rates γ, not of the evolution parameter s. In order to keep the graphical representation simple we have refrained from drawing the homotopy space.

(6) Finally we have to tackle the two basic models from the quantity constrained temporary equilibrium framework from Chapter 7. In doing so we will also fill in the gap left open in Section 12.2. Let us begin with the quantity constrained micromodel from Section 7.1.

Let two points

$$(E_0, e_0) = ((p_0; \widetilde{z}_0^1, \ldots, \widetilde{z}_0^m; F_0^1, \ldots, F_0^m), (\underline{z}_0^1, \overline{z}_0^1; \ldots; \underline{z}_0^m, \overline{z}_0^m))$$
$$\text{and} \quad (E_1, e_1) = ((p_1; \widetilde{z}_1^1, \ldots, \widetilde{z}_1^m; F_1^1, \ldots, F_1^m), (\underline{z}_1^1, \overline{z}_1^1; \ldots; \underline{z}_1^m, \overline{z}_1^m))$$

of the graph of the equilibrium correspondence be given, i.e. two pairs of an economy and an associate equilibrium. Remember that any $\underline{z}_h^a$, $h = 0, 1$, is a point of the compact cuboid $G_- = [-b_1, 0] \times \ldots \times [-b_n, 0] \subset \mathbb{R}_-^n$ and any $\overline{z}_h^a$ is a point of the compact cuboid $G_+ = [0, c_1] \times \ldots \times [0, c_m] \subset \mathbb{R}_+^n$.

In Section 12.2 we have seen that the space of economies is not convex. Nevertheless, as we will see immediately it is fortunately not too hard to take remedial action. As with the other basic models the reader may well conceive of a great variety of alternative ways to achieve this. Here we propose the following intuitive method based on the idea of a twofold composite convex connection evolution.

First remember that

$$\widetilde{z}^a : G_- \times G_+ \longrightarrow \mathbb{R}^n,$$
$$F^a : (G_-)^m \times (G_+)^m \longrightarrow [-b_1, c_1] \times \ldots \times [-b_n, c_n],$$

and that an equilibrium $(\underline{z}^1, \overline{z}^1; \ldots; \underline{z}^m, \overline{z}^m) \in (G_- \times G_+)^m \subset (\mathbb{R}^n \times \mathbb{R}_+^n)^m$ is characterized by the conditions that for every $i = 1, \ldots, n$, and $a = 1, \ldots, m$

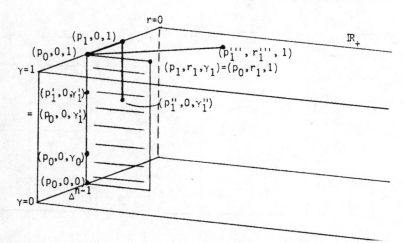

Fig. 11.9: Construction Scheme of a Connecting Economy and Equilibrium Subset for the Basic Model with Production, Taxes, and Subsidies III

(i) $\widetilde{z}_i^a(p; \underline{z}^a, \overline{z}^a) < \underline{z}_i^a \Rightarrow F_i^a(\widetilde{z}^1, \ldots, \widetilde{z}^m) = \underline{z}_i^a$

(ii) $\widetilde{z}_i^a(p; \underline{z}^a, \overline{z}^a) > \overline{z}_i^a \Rightarrow F_i^a(\widetilde{z}^1, \ldots, \widetilde{z}^m) = \overline{z}_i^a$

(iii) $\underline{z}_i^a \leq \widetilde{z}_i^a(p; \underline{z}^a, \overline{z}^a) \leq \overline{z}_i^a \Rightarrow F_i^a(\widetilde{z}^1, \ldots, \widetilde{z}^m) = \widetilde{z}_i^a(p; \underline{z}^a, \overline{z}^a)$.

Before we set out our construction let us illustrate the situation of the pair (E_0, e_0) by Figure 11.10 below. The reader should, however, be well aware of the fact that because of its limitation to two dimensions Figure 11.10 may only be considered as symbolizing the real situation.

Now let us construct a simple auxiliary economy $E_{1/3}$ for E_0 whose equilibrium set contains e_0. Take as $\widetilde{z}_{i_{1/3}}^a$ and as $F_{i_{1/3}}^a$ for every $i = 1, \ldots, n$ and $a = 1, \ldots, m$ the *constant function* with value $F_{i_0}^a(\underline{z}_0^1, \overline{z}_0^1, \ldots, \underline{z}_0^m, \overline{z}_0^m)$. Clearly, $E_{1/3}$ is an economy which serves our purposes.

In complete analogy we construct an auxiliary economy $E_{2/3}$ for E_1.

Now, the reader can reasily convince himself that the composition of the three convex transitions from E_0 to $E_{1/3}$, from $E_{1/3}$ to $E_{2/3}$, and from $E_{2/3}$ to E_1, yields a twofold composite convex connection evolution whose equilibrium set *contains* in each diagram (i, a) of the nm diagrams from Figure 11.10 one of two simple segment patterns.

Clearly, the equilibrium $e_0(e_1)$ is an equilibrium for any s-state economy with $0 \leq s \leq 1/3$ ($2/3 \leq s \leq 1$). Notice further that the equilibrium set of an economy of the simple linear type of our auxiliary economies $E_{1/3}$

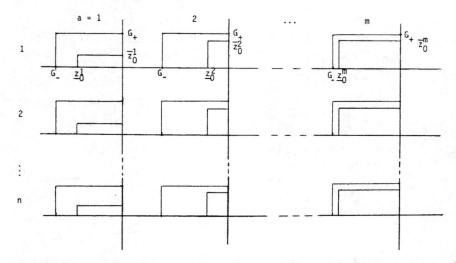

Fig. 11.10: Segment Patterns of a Connecting Economy and Equilibrium Subset for the Basic Models from the Quantity Constrained Temporary Equilibrium Framework

and $E_{2/3}$ in general has a "large" equilibrium set, as the following Figures 11.11 to 11.14 indicate ($\underline{z}_i^a, \overline{z}_i^a$ are from an equilibrium).

(7) Now we come to our last basic model which is left, the basic multisectoral model with quantity constraints from Section 7.2. As before we will cope with the non-convexity of the space of economies by employing a suitable twofold composite convex connection evolution for any two given pairs

$$
(E_0, e_0) = ((l_{1_0}^s(-), l_{1_0}^d(-), y_{1_0}^s(-) y_{1_0}^d(-); \ldots;
$$
$$
(l_{m_0}^s(-), l_{m_0}^d(-), y_{1_m}^s(-), y_{1_m}^d(-));
$$
$$
\widetilde{l}_{1_0}(\widetilde{l}_{2_0}, \ldots, \widetilde{l}_{m_0}; \widetilde{y}_{2_0}, \ldots, \widetilde{y}_{m_0}), \ldots,
$$
$$
\widetilde{l}_{m_0}(\widetilde{l}_{1_0}, \ldots, \widetilde{l}_{m-1_0}; \widetilde{y}_{1_0}, \ldots, \widetilde{y}_{m-1_0});
$$
$$
\widetilde{y}_{1_0}(\widetilde{l}_{2_0}, \ldots, \widetilde{l}_{m_0}; \widetilde{y}_{2_0}, \ldots, \widetilde{y}_{m_0}), \ldots,
$$
$$
\widetilde{y}_{m_0}(\widetilde{l}_{1_0}, \ldots, \widetilde{l}_{m-1_0}; \widetilde{y}_{1_0}, \ldots, \widetilde{y}_{m-1_0}))
$$

and (E_1, e_1) of an economy and an associate equilibrium.

Roughly spoken we first make each given economy "rectangular" such that the given equilibrium is not disturbed. Then we can employ the convex transition between the two auxiliary rectangularized economies. To be more specific, "making rectangular" means for the given economy E_0 to perform the convex transition from E_0 to the following simplified economy with linearized reaction functions and unchanged sectoral equilibria: choose any sector i and consider the parametrizing signal tuple $(l_1, \ldots, \hat{l}_i, \ldots, l_m; y_1, \ldots, \hat{y}_i, \ldots, y_m)$. Now consider the box of the i-th

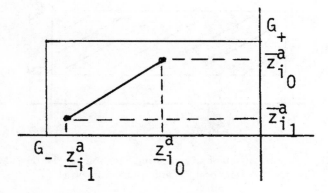

Fig. 11.11: $F_{i_0}^a(e_0)$ and $F_{i_1}^a(e_1)$ Have the Same Sign

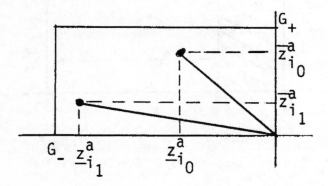

Fig. 11.12: $F_{i_0}^a(e_0)$ and $F_{i_0}^a(e_1)$ Have Different Signs

sector and draw in the "rectangular" crosshair whose intersection point equals the unique sectoral equilibrium $(\bar{l}_i, \bar{y}_i)$. Figure 11.15 below shows the prototypes of the linearized rectangular crosshairs for the nine possible sectoral regimes (cf. Figure 7.4 in Section 7.2). Notice that clearly each rectangular crosshair diagram represents an admissible sector diagram since only reaction functions of the same type, i.e. y^s and l^s, or y^d and l^d, have more than one point in common.

Now deform each reaction function whose graph contains the intersection point $(\bar{l}_i, \bar{y}_i)$ by a convex transition onto the respective aids of the crosshair. (Note that depending on the original sectoral regime between two and four reaction functions are concerned.) The remaining reaction functions (at most 2) are convexly deformed onto the corresponding constant function through the vertex of the original reaction wedge.

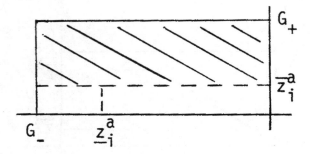

Fig. 11.13: Equilibrium Pattern of a Connecting Economy for the Basic Models from the Quantity Constrained Temporary Equilibrium Framework I

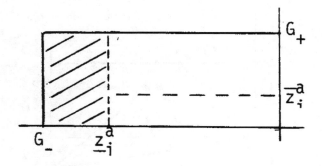

Fig. 11.14: Equilibrium Pattern of a Connecting Economy for the Basic Models from the Quantity Constrained Temporary Equilibrium Framework II

Accordingly, one of the *four types* of rectangular linearized sectoral regimes obtains ("flag patterns") which are shown by Figure 11.16 below. Notice that the described convex deformation leaves the original sectoral equilibrium unchanged. Moreover, for the whole economy, it yields a convex connection evolution whose equilibrium set contains the whole segment $\{e_0\} \times [0,1]$.

The convex transition between the two linearized auxiliary economies finally achieves a connection evolution whose equilibrium set contains a linear equilibrium path from e_0 to e_1. Putting pieces together the twofold composite convex connection evolution from E_0 to E_1, serves all of our purposes. The concluding Figure 11.17 illustrates the nice piecewise linear equilibrium path contained in its equilibrium set in the homotopy space $C^{2m} \times [0,1]$.

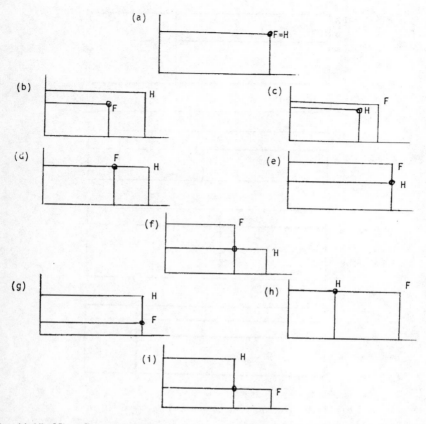

Fig. 11.15: Nine Sectoral Regimes in the Multisectoral Framework with Quantity Constraints

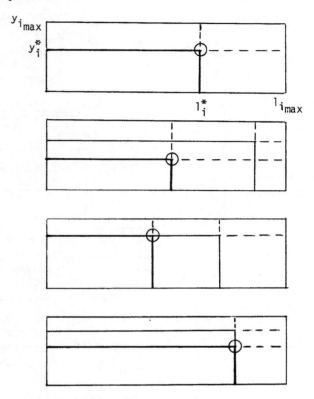

Fig. 11.16: Rectangular Linearized Sectoral Regimes

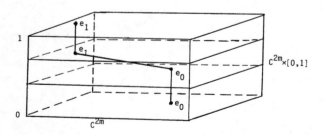

Fig. 11.17: Piecewise Linear Equilibrium Path

Economic Refinements of the Notion of an Evolution of Economies

In this sixth Chapter we will discuss the economic content of our notion of an evolution of economies more thoroughly. In fact, there are two different interpretations of our notion of an evolution of economies possible. On the one hand one may emphasize the *aspect of process*. This means one puts oneself in a position of an observer who at any point of the evolution parameter interval $s \in [0, 1]$ only knows the current state and the preceding states of the evolution which have already been passed by. In other words, the observer "lives through" the exogenously determined course of the varying economic behavior functions. Let us accordingly henceforth call an evolution which is understood in this way a *course economy*.

On the other hand one may emphasize the aspect that an evolution connects its initial state and its terminal state. Accordingly, let us henceforth call an evolution which is perceived in this way a *connection economy*.

The notion of a connection economy immediately raises the central question whether there is *at least one connection economy for any given pair of economies* from any of our basic models. Moreover, it appears to be desirable to have intuitive and standardized constructions achieving this. For instance, the *convex*, or say *linear*, *connection* would be a reasonable candidate. We will see, however, that for some of our basic models the space of economies is *not convex*. But, fortunately, all deviations from convexity will turn out to be curable. More formally this means that we will be able to provide standardized and intuitive auxiliary constructions which may well be accepted as reasonable substitutes for the straightforward convex connection.

Later in this study (Section 11.3) we will demonstrate that these constructions even can be extended so as to achieve simple connecting paths in the graph of the equilibrium correspondence for each basic model type from Part I. Thus we will prove in a completely *constructive way* for each of our basic models that the graph of its equilibrium correspondence is *most nicely connected*. This result significantly extends the well-known result on the manifold property of the graph of the Walras correspondence (cf. Section 9.2). On the other hand, however, we will see that the graph of the equilibrium correspondence exhibits wild geometrical behavior if one looks upon it from arbitrary evolutions (see Chapter 13).

The notion of an evolution of economies in historical time makes it surely of interest also to include the cases of *new commodities* appearing on market for the first time and *old commodities* disappearing from market. In Section 12.2 we will point out that the notion of an evolution as we have introduced it here actually also encompasses these cases in a natural way.

12.1 Course Evolutions and Connection Evolutions

The notion of an evolution of economies as we have introudced it naturally suggests the two different interpretations as an exogeneously determined evolution of economic states, i.e., a *course economy* in progress, and a *connecting evolution* between its initial and its terminal state. To be sure, in our general setting here it makes no difference whether it is an evolution in historical time or an artificial evolution in logical time produced in the economist's laboratory. Summing up, for a *course evolution* one poses oneself in the position of an external observer who 'lives through' the evolution, i.e., who knows at any value $s \in [0, 1]$ of the state parameter only the *current* and the *past states* of the evolution.If one primarily views an evolution under the aspect that it provides a connection of its initial and its terminal state economy we will speak of a *connection economy*.

Let us have a closer look at connection evolutions. A connection economy provides a continuous connection from its initial to its terminal economy which, moreover, satisfies the required properties of an evolution of economies. One also can say that a connection economy provides a continuous path connecting its two border economies in the space of economies topologized with an appropriate topology. In Propositions 4.2 and 5.2 above we have checked this for the usual topology of the space of exchange economies. However, in accordance with our principle of preferring intuitive concepts to abstract ones we will subsequently mainly think of a connection economy as an evolution and not as a path in the space of economies.

Now, two natural questions arise: is there for any basic model from Part I at least one connection economy *for any two given economies*? And is it, moreover, possible to provide general standard constructions achieving connection evolutions which are also economically appealing? The latter particularly means that there should not occur any pathological looking characteristics of the constructed connection economies as they have been discussed in Section 10.1. Clearly, the (piecewise) linear, or straight, i.e. (piecewise) *convex connection,* would meet these requirements notably well. Particularly, it provides a pretty flexible tool with respect to further approximation requirements. The latter is illustrated by the following Figure 12.1. Taking an appropriate

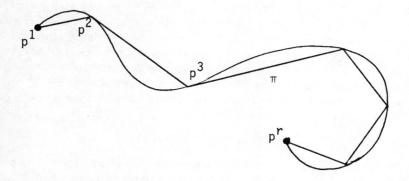

Fig. 12.1: Polygonal Approximation

finite number of intermediate points p^i one can approximate *any continuous path* π in $\mathbb{R}^n$ arbitrarily closely by the composition of the convex connections from each adjacent pair (p^i, p^{i+1}), i.e. by the *polygon* $(p^1 p^2 \ldots p^r)^1$. Actually, we have already made use of the approximating properties of polygons in Euclidean spaces in Section 10.1 when we verified the existence of near-equilibrium paths in the homotopy space.

How the idea of an approximating polygon in Euclidean space can be transferred to evolutions of economies will be the theme of Sections 11.2 and 11.3 below.

Now let us check for which of our basic models the convex connection between any two given economies already is an admissible evolution. What we will find out is that for some basic models the space of economies actually is *not convex.* Nevertheless, the deviations from convexity fortunately will turn out not to be too serious. We will be able to overcome them by providing general intuitive auxiliary constructions which "circumvent" non-convex areas of the space of economies.

---

[1] We remind the reader that the arc as a path's image in general need not give a proper representation of a path, since a path may be a non-one-to-one mapping on its arc. However, throughout this study we only consider one-to-one paths whose arcs give a proper picture of the path itself.

Before we will go through all of our basic models in order to check convexity, and, if necessary, to investigate how to compensate deficiencies we first have to state precisely for any basic model what we understand under a convex *connection*, or *transition*, in the space of economies. Note that we cannot speak of a connection economy before we have not verified that the connection really constitutes an admissible evolution of economies.

Definition 12.1. *Let any two economies ζ^1 and ζ^2 from any basic model from Part I be given. Clearly, for each basic model, except for the model of a large exchange economy (see Section 1.3), ζ^i, $i = 1, 2$, is formally represented by a finite tuple $(f^{i,1}, \ldots, f^{i,k})$ of functions from a Euclidean (sub)space into a Euclidean (sub)space. (For the basic model with quantity constraints and effective demand à la Benassy $f^{i,1}$ is chosen to be constant in $p \in \mathbb{R}_+^n \backslash \{0^n\}$.) The **convex connection**, or **transition**, from ζ^1 to ζ^2 is thus formally given by the one-parametrized k-tuple*

$$(\tau f^{2,1} + (1 - \tau)f^{1,1}, \ldots, \tau f^{2,k} + (1 - \tau)f^{1,k}), \ 0 \leq \tau \leq 1.$$

For the basic model of large exchange economies ζ^i is characterized by a probability measure μ^i on the agents' characteristics space. Accordingly, the convex connection between ζ^1 and ζ^2 is formally given by the one-parametrization

$$\tau\mu^2 + (1 - \tau)\mu^1, \ 0 \leq \tau \leq 1,$$

where $(\tau\mu^2 + (1 - \tau)\mu^1)(B) := \tau\mu^2(B) + (1 - \tau)\mu^1(B)$ for any Borel-set B.

Now we have to check for which basic models convex transitions really form connection evolutions, i.e. admissible evolutions.

Let us start with the basic model of a large exchange economy from Section 4.3. Actually, the only thing we have to make sure is that $\tau\mu^2 + (1 - \tau)\mu^1$ again is a *measure*. This amounts to show that

$$(\tau\mu^2 + (1 - \tau)\mu^1) \left(\bigcup_{l=1}^{\infty} B_l \right) = \sum_{l=1}^{\infty} [\tau\mu^2 + (1 - \tau)\mu^1](B_l)$$

for any sequence (B_l) of disjoint Borel sets whose union again is a Borel set. Clearly,

$$(\tau\mu^2 + (1 - \tau)\mu^1) \left(\bigcup_{l=1}^{\infty} B_l \right) = \tau\mu^2 \left(\bigcup_{l=1}^{\infty} B_l \right) + (1 - \tau)\mu^1 \left(\bigcup_{l=1}^{\infty} B_l \right)$$

$$= \tau \sum_{l=1}^{\infty} \mu^2(B_l) + (1 - \tau) \sum_{l=1}^{\infty} \mu^1(B_l).$$

When $(\tau\mu^2 + (1 - \tau)\mu^1)(\bigcup_{l=1}^{\infty} B_l)$ is *infinite*, then at least one of the two sums must also be infinite, and then clearly the following equations hold:

$$\tau \sum_{l=1}^{\infty} \mu^2(B_l) + (1-\tau) \sum_{l=1}^{\infty} \mu^1(B_l) = \sum_{l=1}^{\infty} \tau\mu^2(B_l) + \sum_{l=1}^{\infty} (1-\tau)\mu^1(B_l)$$
$$= \sum_{l=1}^{\infty} [\tau\mu^2(B_l) + (1-\tau)\mu^1(B_l)]$$
$$= \sum_{l=1}^{\infty} [\tau\mu^2 + (1-\tau)\mu^1](B_l).$$

Correspondingly, when $\tau\mu^2 + (1-\tau)\mu^1 \left(\bigcup\limits_{l=1}^{\infty} B_l \right)$ is *finite*, then both sums $\sum\limits_{l=1}^{\infty} \mu^2(B_l)$ and $\sum\limits_{l=1}^{\infty} \mu^1(B_l)$ also must be finite (convergent), and the chain of equations particularly holds in this case. Thus, we have shown that the space of large exchange economies actually is convex, or in other words, that the convex transition between any two large exchange economies is a connection economy.

Likewise it is straightforward to see that any convex transition in the economy spaces of the two basic Walrasian exchange models from Sections 4.1 and 4.2 is a connection economy. Actually, the Walras property of an intermediate state follows from the linearity of the scalar product

$$p \cdot [\tau\zeta^2(p) + (1-\tau)\zeta^1(p)] = \tau(p \cdot \zeta^2(p)) + (1-\tau)(p \cdot \zeta^1(p)) = 0.$$

Obviously, it is equally straightforward to see that also in the basic models with production, taxes, and subsidies by Kehoe (Sections 3.1, 3.2) any convex transition is a connection economy, i.e. the associated spaces of economies are convex.

Unfortunately, the two basic models from the quantity constrained temporary equilibrium framework from Chapter 7 cannot be settled in the same manner. In fact, both spaces of economies are *not convex*. For the basic quantity constrained micromodel from Section 7.1 one even sees this from a simple example where the voluntariness condition is violated because $\tilde{z}_{i_0}^a, F_{i_0}^a$ and $\tilde{z}_{i_1}^a, F_{i_1}^a$ have different signs for some argument $(z^1, \overline{z}^1; \ldots; \underline{z}^m, \overline{z}^m) \in (G_- \times G_+)^m \subset (\mathbb{R}^n_- \times \mathbb{R}^n_+)^m$. In the basic multi-sectoral model with quantity constraints from Section 7.2 a convex transition easily may generate intermediate sectoral states which are not admissible. The following Figure 12.2 gives a simple example.

Nevertheless, for both basic models the deviation from convexity is not too severe so that it is possible to find reasonable auxiliary constructions overcoming this deficit. However, we will put the reader off to Section 11.3 where we will provide constructions which even admit to control the equilibria in some sense.

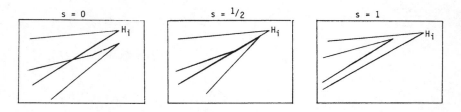

Fig. 12.2: Intermediate Sectoral States of a Convex Transition from the Quantity Constrained Temporary Equilibrium Framework which are not Admissible

Now, there remain the two models from the exchange framework without Walras' law and homogeneity from Chapter 5. Actually, they both stray from the fold. It is not hard to find reasonable examples of convex transitions which do not satisfy the properties required for an evolution. For instance consider the following example which works for both model versions of Chapter 5: let be $n = 2$ and choose the sequence of prices $(p^m) = (m, m)_{m=1,2,\ldots}$. Thus, (p^m) progresses monotonically on the bisector of the positive quadrant of the plane. Now let two basic economies $(\zeta_i^1)_{i=1}^2$ and $(\zeta_i^2)_{i=1}^2$ be given whose excess demand functions have the following shape on a subray of the bisector ($\mathbb{R}_+$ in the following Figure 12.3 symbolizes the subray): Evidently, the 1/2-state

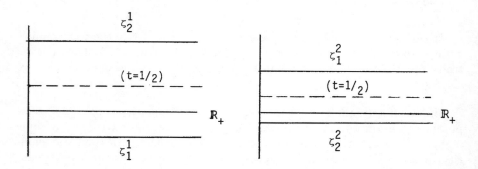

Fig. 12.3: Intermediate States' Scheme of a Convex Transition in the Basic Model Framework without Homogeneity and Walras' Law which are not Admissible

of the convex transition violates property (1) of an evolution in both basic model versions, respectively.

Nevertheless, this example immediately suggests reasonable ways out of this problem. Indeed, the reader may well become aware of a multitude of general constructions which may even be more refined than that we will propose here. Nevertheless, the following construction is simple and intuitive and has the further advantage that it works for both model versions from Chapter 5. Mathematically spoken this construction makes use of the fact that in both models the space of economies is still *star-shaped* even though it is not convex.

To be more specific, let us choose some reasonable economy which qualifies for a 'universal middle state economy' from which *any two* economies can be connected by two straight connections in the space of economies, i.e. convex connection evolutions, as Figure 12.4 suggests: We here propose the following

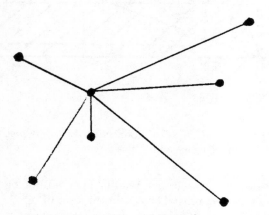

Fig. 12.4: Star-Shaped Connection Scheme for the Basic Model Framework without Homogeneity and Walras' Law

candidate $\zeta^0 = (\zeta_i^0)_{i=1}^n$: choose n excess demand functions ζ_i^0, $i = 1, \ldots, n$, with the following properties:

$(\bar{1})$ There is a real $\alpha > 0$ such that for any $p \in \mathbb{R}_+^n$ with $p_i > \alpha$ and $\frac{p_i}{\sum\limits_{j=1}^n p_j} > \frac{1.1}{10n}$ one has $\zeta_i^0(p) < 0$. (For the second model the condition $\frac{p_i}{\sum\limits_{j=1}^n p_j} > \frac{1.1}{10n}$ clearly can be omitted.)

$(\bar{2})$ For any $p \in \mathbb{R}_+^n$ with $p_j < \frac{1}{20n}$ one has $\zeta_j^0(p) > 0$.

$(\bar{3})$ There is a positive real number b such that for any $p \in \mathbb{R}_+^n$ and any $i \in \{1, \ldots, n\}$ one has
$$\zeta_i^0(p) \geq -b.$$

The following Figure 12.5 illustrates conditions $(\bar{1})$ and $(\bar{2})$ for the first market of an economy with two commodities. Evidently the conditions $(\bar{1})$ and $(\bar{2})$ are so designed that they correspond to properties (1) and (2) of a basic

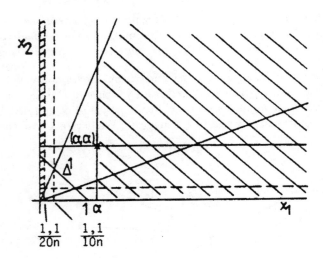

Fig. 12.5: Construction Scheme of a Connecting Economy for the Basic Model Framework without Homogeneity and Walras' Law

economy in our two set-ups without Walras' law and homogeneity and make $(\zeta_i^0)_{i=n}^n$ an admissible economy. The equilibrium set of $(\zeta_i^0)_{i=n}^n$ is contained in $[\frac{1}{20n}, \alpha] \subset \mathbb{R}_+^n$. (This is clear from the construction since $(\alpha, \ldots \alpha)$ lies in the interior of the cone which is spanned by the n vectors v^i with i-th component $v_i^i = 1 - (n-1)\frac{1,1}{10n}$ and $v_j^i = \frac{1}{10n}, j \neq i$.)

Now we have to ensure that any twofold composite convex transition from any economy ζ^1 to ζ^0 and from ζ^0 to any economy ζ^2 actually is a evolution. Formally, the composed, twofold composite convex transition is given by a family $\eta = (\eta_{i_s})_{s \in [0,1]}^{i=1}$ with

$$\eta_{i_s}(p) = \begin{cases} (1-2s)\zeta_i^1(p) + 2s\zeta_i^0(p) & \text{for } s \in [0, 1/2] \\ (2-2s)\zeta_i^0(p) + (2s-1)\zeta_i^1(p) & \text{for } s \in [1/2, 1]. \end{cases}$$

Note that it is clear form properties $(\bar{1})$ to $(\bar{3})$ of ζ^0 that each ζ_i^0, $i = 1, \ldots, n$, can even be chosen so that it equals the $1/2$-state economy of the direct convex transition from ζ^1 to ζ^2, i.e. $1/2\zeta_i^1(p) + 1/2\zeta_i^2(p)$, for any $p \in]\frac{1}{20n} - \epsilon, \alpha - \epsilon[^n$ with an arbitrarily small positive ϵ and an arbitrarily large positive α.

Now we have to verify that for any $s \in [0, 1]$ the s-state mapping $(\eta_{i_s})_{i=1}^n$ actually is an economy. Properties (2) and (3) follow directly from the construction of ζ^0 and η and from the continuity of all functions. As to Property (1) choose any $s \in [0, 1/2]$ and any sequence (p^m) in $\mathbb{R}_+^n \setminus \{0^n\}$ with $\zeta(p^m) \neq \emptyset$. Now choose a $\widetilde{k} \in \zeta(p^m)$ so that $\zeta_{\widetilde{k}}^1(p^m)$ is negative for the arguments of a tail of (p^m). Then, due to the construction, also $\eta_{\widetilde{k}_s}(p^m)$ is negative for the terms of the tail. Clearly, this argumentation directly carries over to the case $s \in [1/2, 1]$. Now we still have to check the uniform boundary condition for n. To see this let (p^m, s^m) be an arbitrary sequence in $\mathbb{R}^n_+ \times [0, 1]$ with $\zeta(p^m) \neq \emptyset$ and $s^m \longrightarrow s^0$. Then at least one of the following statements is true:

(a) there is a subsequence (p^{m_k}, s^{m_k}) in $\mathbb{R}^n_+ \times [0, 1/2]$;

(b) there is a subsequence (p^{m_l}, s^{m_l}) in $\mathbb{R}^n_+ \times [1/2, 1]$.

Together with the above considerations this verifies the boundary conditions, and we are done completely.

12.2 New and Old Commodities

The idea which underlies the notion of an evolution of economic states, be it in real time or just in the economist's laboratory, obviously makes it desirable also to include the two cases of *new commodities* which for the first time appear in the economy, and of *old commodities* disappearing from markets. The question is how this can be incorporated into our set-up such that, moreover, our whole formal treatment can be maintained without major changes. To show that this indeed is possible for each of our basic models will be the purpose of this Section.

Let us start with the group of basic exchange models with finitely many agents from the Sections 4.1, 4.2, 5.1, and 5.2. Consider first an evolution $\zeta = (\zeta_{i_s})^{i=1}_{s \in [0,1]}$ from the Arrow/Hahn or from the Dierker type with $k < n-1$ new commodities which 'enter the picture', i.e. come to market, at $l \leq k$ values of the state parameter $0 < s_1 \leq s_2 \leq \ldots \leq s_k < 1$. This means that there are $n-k$ commodities in the economy when the evolution ζ starts at $s = 0$. At the first glance the requirement of $k < n - 1$ might appear somewhat artificial to the reader who would rather expect $k < n$. In fact, we have chosen this form for technical reasons only. It allows us to make demand functions unaffected by prices of commodities which either are still not, or not anymore, existent in the economy. However, it can be relaxed to $k < n$ for the two models without Walras' law and homogeneity from Chapter 5. We model the market of a new commodity $n - k + j$, $1 \leq j \leq k$, before it enters the picture at s_{n-k+j} in the following simple way. We first choose some arbitrarily small positive ϵ and consider the following canonical compact subspace D^{n-1}_ϵ of the open unit price simplex $\mathring{\Delta}^{n-1} \subset \mathbb{R}^n$ (see Figure 12.6). The compact subspace D^{n-1}_ϵ of $\mathring{\Delta}^{n-1}$ is just $\Delta^{n-1}_{\epsilon/2}$ with the shaded vertex area removed. From now on we consider Δ^{n-1}_ϵ as the economically relevant part of the unit price simplex. We will

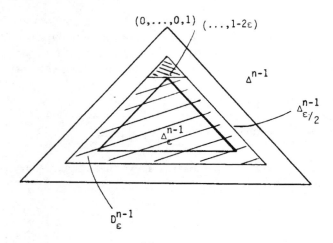

Fig. 12.6: Construction of a Market of a New Commodity

be particularly justified in doing so by excluding all equilibria outside Δ_ϵ^{n-1}. Accordingly the boundary area $\mathring{\Delta}^{n-1}\backslash\Delta_\epsilon^{n-1}$ merely has a technical meaning.

From $s = 0$ to $s = s_{n-k+j}$ any s-state excess demand function of the $(n - k + j)$-th market evolution has the following identical shape: supply and demand are not yet existent and this makes the excess demand function equal to the zero map on the economically relevant part D_ϵ^{n-1} of $\mathring{\Delta}^{n-1}$. It furthermore may have any *zero-free shape* on the boundary area $\mathring{\Delta}^{n-1}\backslash D_\epsilon^{n-1}$ which makes the s-state economy ζ_s an admissible basic economy. From its entry at 'date' $s = s_{n-k+j}$ the graph of the $(n - k + j)$-th excess demand function over D_ϵ^{n-1} may continuously detach from the zero level with growing s.[2] (Clearly, it may also remain further on the zero level when any price vector from D_ϵ^{n-1} also now is an equilibrium price vector for the $(n-k+j)$-th market.) Thus we have achieved our aim.

Evidently, the case of $h < n - 1$ *old commodities* which are going to disappear from market can be treated completely symmetrically. Note that there also might occur the general combined case of commodities which do not yet exist at $s = 0$, and do not any longer exist at $s = 1$. This means they are new commodities which during the evolution become old ones.

Perhaps the reader may wonder at this point why we made those efforts to construct D_ϵ^{n-1} instead of just taking $\Delta_{\epsilon/2}^{n-1}$ as economically relevant part of the price simplex. Actually, so far also $\Delta_{\epsilon/2}^{n-1}$ would well have served for

---

[2] Note that 'date' has been put into quotation marks in order to emphasize that its usual temporal connotation is too narrow for our present considerations.

our purposes. Nevertheless, the construction of D_ϵ^{n-1} will turn out to be most useful for settling the following problem which arises from our approach: what is the economic role of the prices of such commodities which are not any longer, or still not demanded or supplied in the economy? Note that also the prices of commodites which are not existent in the economy during some subinterval of $[0, 1]$ appear as formal arguments of *all* excess demand functions *during* the whole evolution.

To see how D_ϵ^{n-1} can help out of this problem let us fist consider the case of only one new commodity, say the n-th commodity, which enters the scene at $0 < s_n < 1$. Clearly, it would be highly unreasonable when the other excess demand functions $\zeta_{i_s}(p)$, $i = 1, \ldots, n-1$ for $s \in [0, s_n]$ would really be affected by the price p_n of the n-th commodity, that means as long as there is no real demand and supply for this commodity at all. Nevertheless, we will immediately see that one can give p_n the status of a purely technical price argument without any essential effects on the behavior of the agents on the other markets before the entry of commodity n.

To accomplish this let us first notice that any slice D_t^{n-2} of D_ϵ^{n-1} which is parallel to the basis D^{n-2} of D_ϵ^{n-1} and whose points have n-th component equal to t is esssentially the same as D^{n-2} up to a canonical stretching homeomorphism $h_t : D_t^{n-2} \xrightarrow{\approx} D^{n-2}$. This also makes it clear why we chose D_ϵ^{n-1} with the vertex area removed. Now any essential influence of the price of the n-th commodity, p_n, on the other excess demand functions is evidently ruled out if for any two prices p^1 and p^2 from D_ϵ^{n-1} with $p_n^1, p_n^2 \geq 1 - 2\epsilon$ and any $i \in \{1, \ldots, n-1\}$ and $s \in [0, s_n[$ one has

$$\zeta_{i_s}(h_{p_n^1}(p^1)) = \zeta_{i_s}(h_{p_n^2}(p^2)).$$

In words this means that for any s-state market excess demand function ζ_{i_s} with $i = 1, \ldots, n-1$ and $0 \leq s < s_n$ the restrictions to the slices $D_{p_n}^{n-2}$ of Δ_ϵ^{n-2} are essentially identical for $p_n \geq 1 - 2\epsilon$.

Evidently, the case of an old commodity can be treated entirely symmetrically.

It remains now to extend our considerations to the general case of $k < n-1$ new and $h < n - 1$ old commodities. Let us first renumber the commodities such that the first $l \geq 2$ commodities are those which are neither new or old ones. Then follow the h old commodities in the order of their 'dates' of disappearance, and finally the k new commodities in the order of their 'dates' of entry. This means that at state 0 the introduced technique has recursively to be applied k times such that 0-state excess demand on all markets particularly is not affected by the last k prices of commodities not yet existing. When the state parameter s moves on the unaffectedness of excess demand by the last k prices is successively removed at the new commodities' entry dates. On the other hand, in the same recursive way excess demand on all markets becomes successively unaffected by the prices of the old commodities $l + 1, \ldots, l + h$ at the dates of their disappearance. To be sure there is no technical problem with this. Furthermore, the reader should note that the recursive process covers the whole of Δ_ϵ^{n-1}. Therefore it is not necessary, as one possibly might suspect, to choose a smaller $\epsilon_1 < \epsilon$ at the beginning of the process in order to ensure that finally the whole sub-simplex Δ_ϵ^{n-1} is covered.

Reconsidering our constructions the reader will immediately see that the situation with the two models from the exchange framework without Walras' law and homogeneity can even be treated in a simplified way.

So far we have outlined how to incorporate new and old commodities into the introduced formal framework of evolutions basing on the models of pure exchange from Sections 4.1, 4.2, 5.1, 5.2. In a next step let us see how this can be generalized to the basic model 4.3 of a large exchange economy. Also this case is most straightforward. Note that our previous considerations directly carry over to the individual demand functions $f : \mathring{\Delta}^{n-1} \times \mathbb{R}_{++} \longrightarrow \mathbb{R}_+^n$ from the function space D^0. Consequently it seems to be most natural to formalize new and old commodities in this framework by admitting for a certain $s \in [0, 1]$ only such probability measures μ_s, i.e. economies, which assign 0-probability to such individual characteristics $(f, \mathrm{w}) \in D^0 \times (\mathbb{R}_+^n \backslash \{0^n\})$ which do not fit into the specific situation of the economy at state s.

To be sure, probability zero does not mean that a certain unreasonable individual characteristics pair (f, w) may never occur in a model economy μ_s. Nevertheless, what it means is that its contribution to the mean excess demand function of μ_s is zero.

Actually, it is completely straightforward to adapt our technique just provided to the two models with production, taxes, and subsidies by Kehoe (Chapter 3).

Thus, there remain the two models from the quantity constrained temporary equilibrium framework from Chapter 7. Actually, we will see that they are equally easy to treat.

Let us start with the micromodel from Section 7.1. Evidently, since not only the number of commodities is specified by n, but also the number of agents by m, it seems to be desirable to admit changes of both the number of commodities and the number of agents during an evolution in this set-up. Indeed, there is no difficulty to do so. If an agent $j \in \{1, \ldots, n\}$ is not existent, or not active, before $s_j \in]0, 1[$, then his individual *planned* effective demand/supply signals $\widetilde{z}_s^{a_j} = \widetilde{z}_s^{a_j}(p; \underline{z}^{a_j}, \overline{z}^{a_j}) \in \mathbb{R}^n$ just equal the zero vector for $0 \le s \le s_j$. The same applies to his *realized* demand/supply vector $F_s^{a_j}$. The case of an agent who retires from the economy has to be treated symmetrically. (Notice, however, that the continuity assumption for an evolution particularly requires that a retiring agent continuously lowers his economic activities.)

The case of a commodity $c \in \{1, \ldots, n\}$ which enters the scene at $s_c \in]0, 1[$ has to be treated in complete analogy. Up to s_c, i.e., for $0 \leq s \leq s_c$, one has

$$\widetilde{z}_s^a(p; \underline{z}^a, \overline{z}^a) = 0,$$
$$\text{and} \quad F_s^a(\underline{z}^1, \ldots, \underline{z}^m; \overline{z}^1, \ldots, \overline{z}^m) = 0$$

for all $a \in \{1, \ldots, m\}$ and all $(\underline{z}^a; \overline{z}^a) \in G_- \times G_+$. Again, the corresponding case of a commodity which becomes old is completely symmetrical.

The multi-sectoral quantity constrained model from Section 7.2 is likewise easy to treat. When a new sector $j \in \{1, \ldots, m\}$ arises this obviously means that up to some $s_j \in]0, 1[$ there has been no commodity market and no labor market of the j-th sector. In other words, both reaction wedges $F_{j_s}^r$ and $H_{j_s}^r$ of the j-th sector still entirely "lie" in the origin of the j-th box-diagram, i.e. their vertices F_{j_s} and H_{j_s} equal $(0, 0) \in \mathbb{R}^2$ for $0 \leq s \leq s_c$. From s_c on the reaction wedges continuously "move out" of the origin. Again, the corresponding case of a declining sector (a 'sunset industry') which finally disappears from the economy is treated symmetrically. Thus, we have accomplished the aim to incorporate new and old commodities into the formal framework of an evolution for all of our 9 basic models.

13

The Structure of the Equilibrium Price Set of an Evolution of Exchange Economies

As we have already mentioned the existence result of near-equilibrium paths in Chapter 10 leaves us with the following natural question: Is this really the only general structure property of the equilibrium sets of the considered evolutions? And if this is true, how "wild" can a joining equilibrium component of an evolution actually be?

In other words, the question is how complicated the equilibrium set of evolutions can be on the whole. As the reader will remember there is an exhaustive – though disappointing – answer in the literature to the analogue question for the traditional *static* Walrasian exchange set-up. More precisely, after important preparatory work by others Mas-Colell demonstrated in an influential paper from 1977 that *any compact subset* of the price space can be realized as the equilibrium price set of some explicit exchange economy with finitely many consumers characterized by preferences and initial endowments[1] (see Mas-Colell, 1977, Theorem, Corollary 1; for a comprehensive survey on the topic see also Shafer/Sonnenschein, 1982, Section 4). Though being an ingenious contribution in the field of mathematical economics, this result has considerably reinforced the awkward indeterminateness of the exchange framework which had originally been uncovered by the decomposition result by Sonnenschein, Debreu, and Mantel in the early seventies.

We, too, will confine ourselves in this Chapter to the models from the exchange framework(Chapters 4 and 5). We fix our leading question in the following way: *Can any compact subset of the homotopy price space which contains at least one joining connected component be realized as the equilibrium set of some suitable explicit finite exchange evolution, or of some suitable evolution of economies without Walras' law and homogeneity, respectively?* If we can show that the affirmative answer is true, then we will have achieved the

---

[1] Mas-Colell provides a restatement of this result in a differentiable set-up in (1985; Propositions 5.5.8, 5.5.10).

extension of Mas-Colell's static 1977–result to the one-parametrized case. The close assonance of the title of this Chapter to Mas-Colell's 1977–title suggests this. In particular, the affirmative answer means that the existence of at least one near-equilibrium price path in fact is the only general structure property of the equilibrium price set of an exchange evolution.

It has been shown by the author (1988) that the affirmative answer in fact is true for the basic model of an explicit finite exchange evolution (ibid., Section 3, Theorem). Here we will report on this result and its proof, and will generalize it further to our basic set-up without Walras' law and homogeneity.

Following the pattern of Mas-Colell's proof (1977, Proof of the theorem) our proof for the Walrasian exchange set-up will proceed in two stages.

In a **first stage** we will provide a continuous one-parametrization of market excess demand functions whose equilibrium set equals the prescribed compact set. For this we make essential use of an important result in algebraic topological fixed point theory (Schirmer, 1983). In the **second stage** we employ Mas-Colell's decomposing constructions from (1977) to achieve the desired continuous one-parametrization of exchange economies. Actually we will show that the continuous one-parametrization of market excess demand functions achieved in the first step in turn induces *continuous* one-parametrizations of the relevant decomposition constructions by Mas-Colell. (Clearly, the second stage is irrelevant for the basic exchange set-ups of the Dierker type and those without Walras' law and homogeneity.) It is noteworthy that the second stage of our proof, which also proves the first statement of Corollary 13.2 below, fills in a notorious gap in the line of economic justification which is usually given in the literature for this type of decomposition method for excess demand functions. Actually, it has been left open so far in the literature whether the usual decomposition method is continuous, i.e. whether it assigns neighboring explicit exchange economies to neighboring excess demand functions. (To show that the converse is routine, see for instance Mas-Colell, 1985, Prop. 2.7.2.) Clearly, lacking this prerequisitory continuity property any decomposition method appears to be highly artificial and unsatisfactory. The continuity result of decomposition in the differentiable set-up of Mas-Colell (1985, Section 5.8) has been verified in Lehmann-Waffenschmidt (2006).

Now we state our one-parametrized analogue of Mas-Colell's (1977) indeterminateness result on the equilibrium set of an explicit finite exchange economy:

Theorem 13.1. *Let any compact subset K of the one-parametrized price space $S_{++}^{l-1} \times [0,1]$, $l \geq 2$, be given which contains a connected component*

$\widetilde{C}$ *joining bottom and top of the one-parametrized price space, i.e.* $S_{++}^{l-1} \times \{0\}$
and $S_{++}^{l-1} \times \{1\}$.

Then there exists an explicit finite exchange evolution with l *consumers*

$$(E_s)_{s\in[0,1]} : \{1,\ldots,l\} \times [0,1] \longrightarrow \mathcal{P}_{mo}^0 \times \mathbb{R}_{++}^l$$
$$(i,s) \mapsto (\preceq_{i_s}, \omega_{i_s})$$

whose equilibrium set precisely equals the prescribed set K.

The analogue result for the basic model of an exchange economy with finitely many consumers of the Dierker type and for the models from the framework without Walras' law and homogeneity is the following:

Corollary 13.2. *Let any compact subset* K *of any one of the following two one-parametrized price spaces* $\mathring{\Delta}^{n-1} \times [0,1]$ *or* $(\mathbb{R}_+^l \backslash \{0^l\}) \times [0,1]$ *be given which contains a joining connected component* $\widetilde{C}$.

Then for the basic model version by Dierker and for each of the two basic models from the framework without Walras' law and homogeneity from Chapter 5 the following is true: there exists an evolution

$$(\zeta_s)_{s\in[0,1]} : \mathring{\Delta}^{n-1} \times [0,1] \longrightarrow \mathbb{R}^n$$

or

$$(\zeta_s^{I,II})_{s\in[0,1]} : (\mathbb{R}_+^l \backslash \{0^l\}) \times [0,1] \longrightarrow \mathbb{R}^l$$

,respectively, whose equilibrium set equals the prescribed set K.

Being fairly lengthy the proof of Theorem 13.1 will be relegated to Appendix B at the end of the monograph (cf. Lehmann-Waffenschmidt 2006). The Corollary will follow directly from the first part of the proof.

After the statement of the results let us summarize our achievements. First, the results elucidate the structure of the graphs of the equilibrium price correspondences for the addressed basic models of pure exchange. More precisely, they extend Mas-Colell's 1977–result on static economies to the "globalized" case of paths of economies and thus clarify the degree of indeterminateness of the exchange framework on the one-parametrized level.

Together with our earlier findings in Section 11.3 these considerations lead to the following summarizing

Theorem 13.3. *Global characterization of the graph of the Walras correspondence (Chapter 4) and of the graphs of the equilibrium*

price correspondences for the two basic models without Walras' law and homogeneity from Chapter 5:

There is an intuitive standard method to construct for any two points of any one of the addressed three graphs, i.e. for any two pairs (ζ^0, p^0) and (ζ^1, p^1) of an economy and an associate equilibrium price vector, a "well-behaved" path $(\zeta^s, p^s)_{s \in [0,1]}$ in the sense of the constructions in Section 11.3 and in the graph joining these two pairs.

Furthermore, in general nothing more can be said about the equilibrium set of a path $(\zeta_s)_{s \in [0,1]}$ in any one of the considered spaces of economies than that it must be compact and must contain at least one joining equilibrium component admitting a well-behaved (near-)equilibrium price path.

More informally speaking, our findings mean that any of the three considered graphs is "nicely connected". But it is still complex enough to admit any wild configuration of equilibrium prices as long as it is compact and contains a joining equilibrium component generated by some economy of economies in the respective model.

For a detailed discussion of the relationship between our results and the well-known structure results on the graph of the Walras correspondence from the literature the reader is referred to Section 14.1.

Secondly, there are, nevertheless, still further gains from our results. Particularly, they extend an important result by B. Allen (1981, Theorem 5.1, Corollary 5.3) in a certain respect. Let us first briefly recapitulate this result. Theorem 5.1 by B. Allen says for the Walrasian exchange framework that for any given *smooth one-to-one selection* from the graph of the Walras correspondence over some evolution of economies forming a smooth finite dimensional manifold and for any *sufficiently close smooth* function into the graph one can realize also this latter function as a selection from the graph over some manifold of economies being *close* to the original one. An immediate implication of this is that any vector from the price simplex which is sufficiently close to some arbitrary equilibrium price vector of some economy actually is an equilibrium price vector of some *neighboring* economy, which, moreover, can even be chosen from some sufficiently rich subset of an open and arbitrarily small neighborhood of the original economy (Allen (1981), Corollary 5.3).

Let us now discuss the relationship of B. Allen's reported results to our results from this Chapter. First, we note the following striking consequence of Theorem 13.1.

Proposition 13.4. *For any one of the four spaces of explicit finite exchange economies (see p. 22), of exchange economies of the Dierker type (Section*

4.2), or of the two types of exchange economies without Walras' law and homogeneity (Section 4.3) the following holds:

let any two arbitrary non-empty compact subsets of the price space be given, then these sets can be realized as the equilibrium sets of two arbitrarily close economies.

Proposition 13.4 is proven in Appendix B.

This result reinforces the indeterminateness of the static Walrasian exchange model which was detected by the decomposition result by Sonnenschein/Debreu/Mantel. To paraphrase Proposition 13.4 it says that looking all around the graphs of the considered equilibrium correspondences one always finds somewhere two arbitrarily close economies whose equilibrium sets equal two arbitrarily prescribed compact sets. The reader should well notice, however, that this result provides no further information on the position of these economies in the space of economies. Particularly, this means that our results neither recover, nor extend what B. Allen rightly characterizes as her "localized analogue of Mas-Colell's (1977) result on the lack of restrictions on the equilibrium price set". However, in the other respect of "which equilibrium sets can be generated at all by evolutions of economies?" our results go beyond the cited results by B. Allen. To be more specific, restricting B. Allen's Theorem 5.1 to the case of paths of economies our Theorem 13.1 is, apart from relaxing differentiability requirements, more general in the following way: we are not confined to continuous *one-to-one selections* from the graph of the Walras correspondence which, moreover, even must be close to some *given* one-to-one selection. Instead, we may admit *any compact subset K* of the homotopy space containing a joining connected component as a candidate for the equilibrium set of some evolution of economies.

Comparison With Related Results in the Literature

At the end of Part II of our study we now will point out the precise relationship of our approach and results presented so far to related work from the literature. We, however, would like to emphasise that this only concerns the formal results, whereas any further economic applications are new and can be found in Part III of our study.

In Section 14.1 we draw a comparison between our approach and results and the studies of the graph of the Walras correspondence mainly developed and advanced by Y. Balasko. In Section 14.2 then we will point to the gains and losses of our analysis in relation to the well-known theory of regular economies which has been initiated by Debreu in his seminal paper from 1970. A comprehensive presentation of the achievements of the regular approach has been given by Mas-Colell in his book from 1985. Especially the results which Mas-Colell provides in the Sections 5.8 and 8.8 are closely related to ours. Nevertheless, as we will point out there are significant advantages of our approach and results.

14.1 Studies of the Graph of the Walras Correspondence

Since the seventies the structure of the graph of the Walras correspondence has been studied in a series of papers, mainly contributed by Y. Balasko (see Balasko's contributions in the reference list, particularly his survey monograph on the topic from 1988, and the surveys by Dierker (1982), and Mas-Colell (1985, Sections 5.8, 8.8)). In order to make clear the differences between these results and ours presented so far we first briefly report on the approach and the results of this research branch. As before we will keep as close as possible to the original notations.

There are n commodities and m agents each of whom is characterized by a *fixed* C^r-demand function ($r \in \{0, 1, 2, \ldots\}$)

$$f^i : \mathring{\Delta}^{n-1} \times \mathbb{R}_{++} \longrightarrow \mathbb{R}_+^n, \quad i = 1, \ldots, m,$$

satisfying Walras' law

$$\forall p \in \mathring{\Delta}^{n-1} \ \forall w_i \in \mathbb{R}_{++} \ p \cdot f^i(p, w_i) = w_i.$$

Denoting by $P^m = \mathbb{R}_+^{nm}$ the space of admissible initial endowment tuples $(\omega^1, \ldots, \omega^m)$ of the m agents and by f the m-tuple $(f^1, \ldots, f^m)$ the primary concern is with the global properties of the *graph* E_f of the *Walras correspondence*

$$W : P^m \longrightarrow \mathring{\Delta}^{n-1}$$
$$(\omega^1, \ldots, \omega^m) \mapsto \{p \in \mathring{\Delta}^{n-1} | \sum_{i=1}^m f^i(p, p \cdot \omega) = \sum_{i=1}^m \omega^i\}.$$

(Of course, the projection of E_f on the space of economies P^m is not onto, since there are economies $(\omega^1, \ldots, \omega^m)$ which possess no equilibrium price vectors.)

Summarizing the results from the literature E_f is path connected and simply connected. In fact, it is a contractible C^r-manifold and, moreover, it is C^r-diffeomorphic to the Euclidean space $\mathbb{R}^{nm}$ (Balasko (1975 b), (1988), Schecter (1979), see particularly also Mas-Colell(1985), Propositions 5.8.22 and 8.8.1 for a survey).

Furthermore, the analysis can be extended to the case of *varying demand functions*. Topologizing the set of all demand functions by the compact open topology and denoting it by K the following is true (Balasko (1975a)): the union of the equilibrium manifolds associated with a path connected subspace $G \subset K$, i.e., $\bigcup_{f \in G} E_f$, is a path connected subspace of the Euclidean subspace $P^m \times \mathring{\Delta}^{n-1}$. In particular, the whole space K of demand functions is path connected (since it is convex). Hence, the whole union $\bigcup_{f \in K} E_f$ is path connected. Therefore any two pairs $(\zeta_1, p^1), (\zeta_2, p^2)$ of economies $\zeta_j = (\omega_{1j}, \ldots, \omega_{mj}; f_{1j}, \ldots, f_{mj}), \quad (j = 1, 2)$, and associate equilibrium price vectors p^j (i.e. for $j = 1, 2$ the equation $\sum_{i=1}^n f_{ij}(p^j, p^j \cdot \omega_{ij}) - \sum_{i=1}^n \omega_{ij} = 0$ holds) in principle can be connected by a continuous path (ζ_t, p^t) of such pairs in the product space $\mathcal{E}' \times \mathring{\Delta}^{n-1}$. (The space of economies $\mathcal{E}'$ is topologized by the topology of C^0-uniform convergence on compact subsets.) However, there is no further information on the connecting path (ζ_t, p^t) available. Particularly this means that neither the economy ζ_t nor the equilibrium prices p^t are known for $0 < t < 1$.

Having reported on the well-known studies of the graph of the Walras correspondence we have to point out what the advantages of our results are.

Besides the fact that we are not confined to the single basic model of pure exchange there are two major advantages.

(1) In contrast to the reported approach we examine the changes of the equilibrium set over arbitrary and definite paths of economies $(\zeta_s)_{s \in [0,1]}$ which we call evolutions of economies. Actually, the existence of a (possibly backtracking) (near-)equilibrium price path $(p^s)_{s \in [0,1]}$ for (ζ_s) cannot be derived alone from the contractible manifold property of the graph of the Walras correspondence as the following counterexample shows: a two-dimensional contractible submanifold in $\mathbb{R}^3$ can for instance be twice folded-over as in the following Figure 14.1 (consider the $x_1 x_2$-coordinate plane as space of economies $\mathcal{E}'$). Consider the evolution of economies which

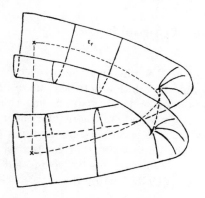

Fig. 14.1: Two-Folded Equilibrium Manifold

is given by the straight line segment in $\mathcal{E}'$. Evidently the part of the graph of the Walras correspondence lying above it consists of *two separate pieces* (cf. the "cross-section slice" in the following Figure 14.2).

(2) In Section 11.3 for any of our basic models we have provided an economically appealing standard method how to provide a simple path connecting any two points of the equilibrium correspondence *within* the graph of the equilibrium correspondence. This not only gives a *constructive proof* of the path connectedness of the graph of the equilibrium correspondence, but over and above that it proves that the graph is even *most nicely* path connected. This is clearly not implied by Balasko's abstract result of the manifold property of the graph of Walras correspondence.

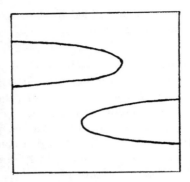

Fig. 14.2: Cross-Section Slice of the Two-Folded Equilibrium Manifold

14.2 The Regular Approach

After we have made lucid the differences between our results and those by Balasko we will point out the relation between our results and the influential theory of regular economies. Actually, this means two tasks. First we will report and discuss the original body of the regular theory *for static economies* as it has been initiated by Debreu in his seminal paper from 1970. Second we will report and discuss the important generalization of the regular approach to *one-parametrized paths of economies* by Mas-Colell in his comprehensive monograph from 1985 (especially Section 8.8).

Let us start with the first issue.

The theory of regular economies (see e.g. Debreu (1970, 1976), Dierker (1974, 1982), Mas-Colell (1985)) works with the usual exchange model: there are n commodities and m agents. $\mathcal{D}^i$ denotes the space of m-tuples of individual C^i-demand functions

$$f^j : \overset{\circ}{\Delta}{}^{n-1} \times \mathbb{R}_{++} \longrightarrow \mathbb{R}^n_+, \quad j = 1, \ldots, m,$$

topologized by the topology of uniform C^i-convergence on compacta, where the demand function of at least one agent has the well-known desirability characteristics: if (p^k, w^k) is a sequence in $\overset{\circ}{\Delta}{}^{n-1} \times \mathbb{R}_{++}$ which converges to $(p^0, w^0) \in \partial \overline{\Delta}^{n-1} \times \mathbb{R}_{++}$, then there is at least one index $j \in \{1, \ldots, m\}$ so that $\|f^j(p^k, w^k)\|$ grows beyond all finite bounds. $P^m := (\mathbb{R}^n_{++})^m$ denotes the space of initial endowment-n-tuples. Any *economy* $(f^1, \ldots, f^m, \omega^1, \ldots, \omega^m) \in \mathcal{D}^0 \times P^m$ satisfying Walras' law

$$\forall_{p \in \overset{\circ}{\Delta}{}^{n-1}} \quad p \cdot \left(\sum_{j=1}^m f^j(p, \omega^j \cdot p) \right) = p \cdot \left(\sum_{j=1}^m \omega^j \right)$$

has equilibrium price tuples, i.e. the set of all price tuples p with

$$\sum_{j=1}^{m} f^j(p, \omega^j \cdot p) = \sum_{j=1}^{m} \omega^j$$

is non-empty (see e.g. Dierker (1974), Theorem 8.3). The Walras correspondence $W : \mathcal{D}^0 \times P^m \longrightarrow \mathring{\Delta}^{n-1}$ assigns to every economy $(f^1, \ldots, f^m, \omega^1, \ldots, \omega^m)$ its equilibrium set.

For $\mathcal{D}^0 \times P^m$ one can show that W is upper hemi-continuous (Dierker (1974), Theorem 8.4). This means the equilibrium set cannot 'explode' by slight variations of the data. But it can 'implode'. More precisely, at 'critical' economies certain subsets of the equilibrium set may suddenly disappear, even if one only permits arbitrarily slight variations of the data. However, at the economies of a *residual subset* of $\mathcal{D}^0 \times P^m$ neither explosions nor implosions of the equilibrium set may occur if data change appropriately slightly (Dierker (1974), Theorem 8.5, Corollary). Furthermore, the subset of economies with finite equilibrium sets is even dense (Dierker and Dierker (1972)).

The *theory of regular economies* deals with $\mathcal{D}^1 \times P^m$. A regular economy has a finite equilibrium set, the number of equilibria is odd, and, furthermore, is locally constant with respect to varying data (e.g. Debreu (1970,1976), Dierker (1972, 1982), Dierker and Dierker (1972)). Appropriately small variations of the initial endowment tuples, or of the demand functions, lead to continuous and disjoint movements of the finitely many equilibrium price tuples in the price simplex. The main result of the theory is that the subspace $\mathcal{R}$ of regular economies is not only a dense, but even an *open* subspace of the whole space $\mathcal{D}^1 \times P^m$. In fact, the projection mapping $\mathcal{R} \times \Delta^{n-1} \longrightarrow \mathcal{R}$ is a covering map (Dierker and Dierker (1972)).

Thus, the theory of regular economies provides a *local, strong result*: if the data of a regular economy E vary, the movements of the finitely many equilibrium price tuples produce continuous traces in the price simplex as long as the data do not leave that connected component of $\mathcal{D}^1 \times P^m$ which contains E.

After this brief report on the regular approach and its achievements in the static framework we are now going to discuss it against the background of our analysis. The most striking feature of the regular approach is that it just excludes all critical economies from the analysis. But it are precisely the *critical economies* where the important and interesting changes of the equilibrium set take place. More formally, no information at all is provided by the regular theory when one leaves a connected component of the economy space $\mathcal{D}^1 \times P^m$. This naturally makes one curious about how many connected regular components do exist and how complex the subdivision is by connected components of the space of economies. As to the first question the answer is immediately clear from the definition of a regular component: there are at least as many regular components as there are natural numbers, i.e. numbers of equilibria. As to the second question the meaningfulness of the regular theory surely would be the greater the more simply, or say regularly, the regular components subdivide the space of economies.

Before we start with analytically tackling this question let us support intuition by the following *two examples* of a subdivision of the plane. Obviously, points like A in the "winding spider's web" in Figure 14.4 below make this subdivision of the plane much more complex than in the example of the "regular grid" in Figure 14.3. Indeed, arbitrarily slightly different movements away from A in Figure 14.4 may lead into eight different components each of which exhibits unpredictable properties. Unfortunately, we will find that it is rather

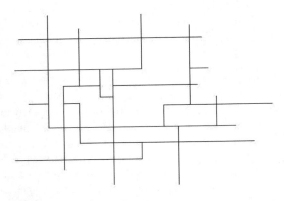

Fig. 14.3: Regular Grid Subdivision of the State Space

the winding spider's web than the regular grid which adequately illustrates the situation with the subdivision of the space of economies by its regular components.

In fact, the reader can easily convince himself that there are critical economies which correspond to points like A in Figure 14.4. He just has to perform different slight perturbations of the simple one-dimensional critical excess demand function $\zeta^{1/2}$ of Figure 14.5. Moreover, regular components in general are not convex. This is symbolized by the winding lines of the web in Figure 14.4 below. Actually, Figure 14.5 shows an example of a convex transition from ζ^0 to ζ^1 generating the critical 1/2-excess demand function $\zeta^{1/2}$. Figure 14.6 transforms this situation into the non-convex subdividable state space from Figure 14.4. A reader who is keen on experimenting with geometrical

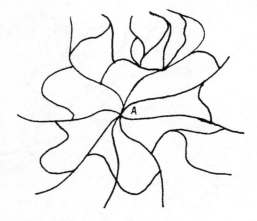

Fig. 14.4: Non-Convex Subdivision of the State Space

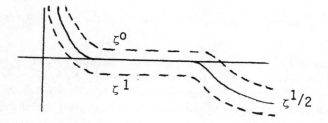

Fig. 14.5: Convex Transition with Critical Mean State in a Functional Diagram

examples easily will find a great variety of other situations which even tend to complicate the scene further[1].

Having discussed now the gains and losses of the well-known regular approach at some length let us come back to our approach. Besides the fact that we do not stick to the traditional exchange framework our analysis has the major advantage that it is not confined to a fixed regular component. In fact, our evolutions may run anywhere in the space of economies. Particularly, they not only may connect any two regular economies from any two regular components, but also even any arbitrary pair of critical economies.

The primary economic consequence of our observations is that the label 'regular' must be considered critically. Actually, if we are given a certain regular economy the regular theory says nothing to us what happens if we perturb

---

[1] In Section 17.1 below we will see in a different context that the subdivision of the regular subspace by non-regular economies is even still more complex than our present considerations show.

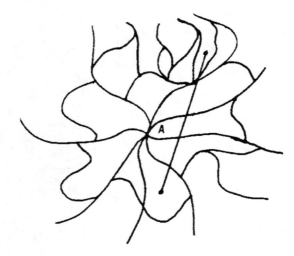

Fig. 14.6: Convex Transition with Critical Mean State in the State Space

it to the boundary of the regular component containing it, or beyond. More-over, if we are given a certain non-regular economy there is evidently great arbitrariness of which regular component we reach even when we only slightly perturb. Leaving the propedeutical examples with merely one dimension con-fusion is even increased.

There is still another argument frequently used to justify the regular ap-proach. This is expressed by the lable 'generic' or 'typical' for regular econo-mies, and 'untypical', 'exceptional', or 'negligible' for non-regular, i.e. critical, economies.

The original motivation of this characterization is clear: the set of non-regular economies actually is *small* from the *topological* and *measure theoretical* viewpoint. However, the question remains whether smallness from a purely mathematical viewpoint really justifies it to consider non-regular economies as negligible. In defense of the negligibility characterization the argument is usually advanced that "nature does not act on null sets". Let us have a closer look to this kind of argument. It evidently relies on the implicit hypothesis that — if one accepts the exchange framework as a reasonable description of real economic systems — any economy from the whole space of economies has the same probability of being realized (uniform probability distribution). But, how is this hypothesis justified? Why should nature not prefer an unequal probability distribution with even higher probabilities in regions where critical states concentrate? What does justify the confidence in critical states having zero probability? Certainly, the practical experience with real economic systems seems not to support such an optimistic confidence.

So far we have discussed the main body of the 'classical' theory of *static* regular economies. Now we come to the second item drawn up at the outset of this Section. In his influential book (1985) Mas-Colell has generalized the traditional static regular approach to the analysis of the generic properties of the graph of the Walras correspondence over paths of economies, i.e. over evolutions (ibid., Section 8.8). The reader should note that here Mas-Colell confines his analysis to the basic model of an explicit *finite* exchange economy (see ibid., p. 344).

In the central Proposition 8.8.2 Mas-Colell shows the analogue of the main result of the classical regular theory for *smooth paths of economies*: the "regular" paths of economies are *generic*, i.e. form an *open and dense* subspace J' of the space J of smooth paths of economies. For our purposes it is sufficient to notice that a *regular path of economies* particularly has a *regular* initial and terminal economy. Moreover, it has an equilibrium set which is, up to diffeomorphism, a finite disjoint union of circles and segments, the endpoints of the segments precisely equalling the intersection of the equilibrium set with $\Delta^{n-1} \times \{0\}$ and $\Delta^{n-1} \times \{1\}$ (cf. Mas-Colell, 1985, p. 344). In a further result Mas-Colell establishes genericity of those paths of economies whose equilibrium sets nicely project on the price space Δ^{n-1} of the homotopy prism $\Delta^{n-1} \times [0,1]$. More specifically:

Proposition 14.1. (Proposition 8.8.5, Mas-Colell 1985): *If there are at least 4 commodities, then there is an open dense subspace J'' of the space J of C^∞-paths of economies such that every $\eta \in J''$ is a regular path and the projection of its equilibrium set $E_\eta \subset \Delta^{n-1} \times [0,1]$ on the price simplex is one-to-one.*

In other words, given any path $\eta = (\eta_s)_{s \in [0,1]} \in J''$ any price vector from Δ^{n-1} can be an equilibrium price vector for at most one s-state economy η_s, $0 \le s \le 1$. This is symbolized by the following Figure 14.7. Indeed, the one-to-one projection on the price space Δ^1 is a fairly restrictive property, and consequently Proposition 8.8.5 appears to be an astonishing result. The reader should, however, be well aware that Proposition 14.1 actually *does not work* for the case depicted in Figure 14.7 where the number of commodities is only two. Actually, in higher dimensions the result of Proposition 14.1 appears to be more restricted.

Knowing the result of Proposition 14.1 one naturally may ask whether there is an analogous result with the *second projection* on the interval $[0,1]$ of state parameters. To be sure such a result would mean that all evolutions which contain at least one state economy with a non-unique equilibrium set

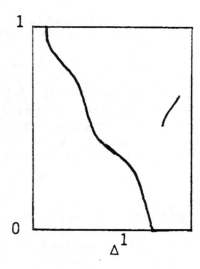

Fig. 14.7: Illustration of Proposition 14.1

– particularly all evolutions with a backtracking (near-)-equilibrium path –
would form a closed null set in the space of evolutions. Unfortunately, there is
no such result. Instead, Mas-Colell provides the following significantly weaker
result in this direction:

Proposition 14.2. (Proposition 8.8.3, Mas-Colell 1985): *There is an
open and dense subspace $J''' \subset J$ such that every path from J''' is regular and
the projection of its equilibrium set on $[0,1] \subset \mathbb{R}$ is a Morse function.*

Let us recall that a C^2-function $f : M \longrightarrow \mathbb{R}$ from a manifold M into
the reals is a *Morse function* iff *all its critical points,* i.e. where $\partial f(x) =
0$, are *nondegenerate,* that means satisfy a certain regularity condition with
respect to the second derivation (see Mas-Colell, 1985, p. 39). Particularly,
however, the Morse property does not prevent the occurrence of infinitely
many critical points or of *backtracking* joining equilibrium paths. The following
Figure 14.8 gives an illustration (cf. Figure H.2.3 in Mas-Colell, 1985): x is a
nondegenerate critical point, whereas x' is not.

The nice properties of the equilibrium sets of regular paths of exchange
economies naturally lend themselves to an application in the field of *computation of equilibria.* Actually, following a joining segment in the equilibrium set of
a regular path which starts at a well-understood regular initial economy and
ends at some terminal regular economy whose equilibrium set is unknown provides an explicit procedure for computing an equilibrium from the unknown

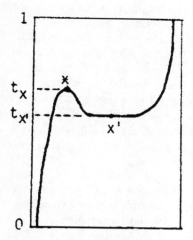

Fig. 14.8: Critical and Non-Critical Points Consistent with Morse Property

equilibrium set. Mas-Colell (1985, p. 207) provides a certain regular economy from which the convex path, i.e the convex connection evolution, to any given regular economy is regular. Moreover, there is even an explicit *path following algorithm* for following the unique joining equilibrium segment. It is based on an implicit system of differential equations (ibid., p. 209). There is, however, the severe limitation of this method that it only can detect equilibria with index +1 (see Mas-Colell, 1985, pp. 209–211).[2]

After this report on the virtues of the one-parametrized regular approach in the basic framework of explicit finite exchange economies let us point out where our results have advantages. The reader certainly will have noticed that the reported open-density result on regular paths, i.e. Proposition 8.8.2 by Mas-Colell, is closely related to our analysis of the Sections 11.1 and 11.2.

---

[2] The reader can find a comprehensive survey on the theory and applications of the path following method for computing zeroes of functions for instance in Leininger (1978). Section 5.3 of Leininger's study presents the theoretical background of a path following algorithm based on a system of differential equations. Chapter VI provides a practical numerical realization of the path following algorithm. Furthermore, in Section 5.4.1 it is demonstrated (Satz 5.4.1) that the path following method in fact is equivalent to the method introduced by S. Smale in his celebrated paper from 1976 (cf. also Mas-Colell, 1985, pp. 211–214). We will come back to the path following method in Section 10.1. We further refer the interested reader to the contributions on equilibrium computations for instance by Manne, Talman and van der Laan, and W.C. Rheinboldt. The last author particularly stands for the growing branch of equilibrium computation in numerical mathematics (see the reference list at the end of the monograph).

There we have constructed approximating evolutions with nice joining equilibrium paths for any one of our basic set-ups. Now, what are the gains of our method compared to Proposition 8.8.2 by Mas-Colell? *First*, we provide *explicit constructions* whereas Proposition 8.8.2 just is an abstract existence result which gives no further information on the specific shape of approximating paths. To be more specific: perturbing a critical path of economies only leads with probability 1 to a regular one. Furthermore, its properties extremely vary with the chosen direction of the perturbation. And *second*, our construction principles can directly be generalized to our other basic models, particularly to the basic model of a large exchange economy. *Finally*, we need no differentiability assumptions.

To be sure, the equilibrium set of an approximating *regular* path is in general in one respect more well-behaved than the equilibrium set of an approximating evolution achieved by our methods in Chapter 11 of our study, but it is less well-behaved in another respect. This means, the equilibrium set of a regular path is up to diffeomorphisms a finite union of segments and circles. But, diffeomorphisms still allow for geometrically bad behaviour like for instance infinitely many oscillations and infinite length for joining equilibrium segments.

Let us sum up the weaknesses of both approaches: Mas-Colell's result still allows for unpleasant features from the viewpoint of geometrical intuition, whereas in our set-up indeterminateness like bifurcations or "thick" parts of joining equilibrium components may still occur. Moreover, Mas-Colell's result of a regular approximating path does not automatically admit an explicit path following algorithm for following one of the joining equilibrium segments. This is for the simple reason that Proposition 8.8.2 only ensures existence of regular approximating paths, but does not specify them. Furthermore, in general the regular path following method does not permit to approach equilibria of *critical* economies. This, in contrast, is made possible by our method – at the price of missing a general explicit path following algorithm (cf. Section 17.1 below).

We now conclude our discussion by an important observation on the structure of the *subspace of regular paths*. Actually, the following result will show in complete analogy to the static case that also the space of smooth paths of Walrasian exchange economies J' is subdivided by its regular components in a way which is at least as complex as the winding web of Figure 14.4 above illustrates. Indeed, Proposition 14.3 gives the basis for our result that there are infinitely many components of the subspace of regular evolutions (paths) of economies:

Proposition 14.3. *The mapping* $\psi : J' \longrightarrow \mathbb{N}$ *which assigns to every regular path the number of its joining equilibrium components is continuous.*

Proposition 14.3 immediately implies

Corollary 14.4. *There are at least as many regular components as natural numbers.*

The proof of Proposition 14.3 is straightforward and geometrically intuitive. Being lengthy, however, we will relegate it to Appendix C.

As for the static case we may easily convince ourselves that the winding web from Figure 14.4 is an adequate illustration by experimenting with a simple one-dimensional example. Figure 14.9 shows a non-regular evolution (a) which generates a 'pitchfork' equilibrium set (Figure 14.9a'). Arbitrarily small variations of the evolution (a) to regular evolutions (b) or (c) lead to completely different equilibrium sets ((b') or (c')). This finishes our description of the complex structure of the subspace of regular paths of Walrasian exchange economies.

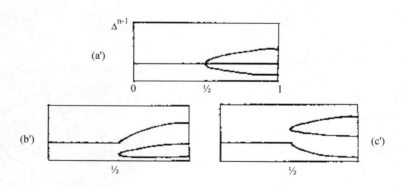

Fig. 14.9: Patterns of Evolving Equilibrium Sets and their Generating Functional Evolutions

15

Conclusions

Part II of the study contains the analytical results on economic evolutions that were conceptualized in Part I. Exploiting the unifying constructions in Part I, we have shown in the present part of the book that any admissible evolution possesses a geometrically well-behaved equilibrium path, or at least a well-behaved approximating path in the equilibrium correspondence. Using an approach tht is completely different from the one used here, A. Mas-Colell (1985) has independently shown this result for the basic set-up of a large exchange economy. In addition, we have given a sufficient algebraic criterion for identifying points on equilibrium paths for any given evolution. Both results heavily rely on results from one-parametrized algebraic fixed-point theory.

In Chapter 11, we turn to the question of whether any evolution can be approximated in the space of economies such that an evolution is obtained, which has a well-behaved equilibrium path. This question has also been tackled in a different context by A. Mas-Colell for the basic model of an exchange economy.

The advantages of our method are the following ones: We give constructive methods for achieving approximating evolutions, whereas Mas-Colell merely provides an abstract existence result. Our constructions particularly allow for an extension to all of the basic set-ups introduced in Part I. The second advantage is that we can derive further standardized constructions that show that for each of the basic models, the graph of its equilibrium correspondence is extremely well-connected.

In Chapter 12, we outline the two different interpretations of an evolution as a course evolution on the one hand, and as a connection evolution on the other hand. In particularly we provide standard constructions that show that for any of the basic set-ups there is actually a connection evolution for *any* two given economies. Furthermore, we show how to extend the formalizations of evolutions given in Part I to also include the cases of new commodities entering the markets, and old commodities leaving them during an evolution.

In Chapter 13, we give an answer to the question of whether our general structural property of the existence of (near-) equilibrium paths for *exchange* evolutions is really the only one which generally holds. In fact, the answer is yes. This means we have achieved the one-parametrized analogue of Mas-Colell's famous 1977 result on the unrestrictedness of the equilibrium set of static exchange economies. Together with B. Allen's (1981) 'localized' analogue of the addressed 1977 result by Mas-Colell, our result achieves a fairly comprehensive understanding of the equilibrium sets of evolutions of exchange economies.

In Chapter 14, we finally make precise the relationships between our approach and results and the theory of regular economies and the studies of the graph of the Walras correspondence by Y. Balasko and others. Besides placing one-parametrizations of economies in a more thorough economic context, our results, in a number of aspects, significantly extend those by the two established branches of equilibrium theory addressed.

Economic Analysis

16

Introduction to Part III

In Part III of our study, we will explore the economic content of the concepts from Part I and of the analytical results derived in Part II.

Following the usual classification, we will provide atemporal and temporal applications. In the *atemporal* realm, the results lead to extensions of two well-established techniques of economic theory. More specifically, we are able to provide a certain kind of an extension of the so-called path following method to also compute equilibrium points of non-regular economies. Second, based on the results presented in Part II we will provide a new way for how to give comparative statics an economic meaningfulness even in the case of multiple equilibria.

When speaking of a *temporal* economic analysis one usually thinks of a dynamic analysis. This, however, will not be the approach in this study when dealing with temporal applications. We will confine ourselves to what we call a 'kinetic analysis'. This means we do not attempt to explain the causal relationships between successive states of an economic evolution, but strive towards a general analysis of the effects of the evolution of the economy on the endogenous solution (=equilibrium) values. This will be the focus of the last three chapters of the study.

In Chapter 18, we will introduce and discuss the kinetic method at some length. In Chapter 19 we will conceptualize evolving economies in discrete and continuous historical time and will apply our results from Part II. The major focus of our applications, however, is on evolving economies in continuous historical time (Sections 19.2, 19.3 and 19.4). We will introduce two alternative models of evolving economies in continuous historical time (Section 19.2). The first one, which we will call the 'flow commodity model', is in some sense inspired by continuous growth models – however, without any use of differential equations. The second one, which we will call the 'frequency model', is completely new.

Generally speaking, the results state that it is possible for any evolution of the economic system which is not influenced by any policy agency to permanently readjust equilibrium values in a continuous way – generally up to finitely many dates at which discontinuous jumps are unavoidable. The economic gains of such a piecewise continuous adjustment over time of equilibrium variables are best understood when thinking of discontinuous changes. First, discontinuous changes of equilibrium values make it hard to find new equilibria of new states of an evolving economy. Moreover, discontinuous changes not only entail sudden and abrupt changes of the economic agents' status and behavior, but also disturb and destabilize the agents' expectations. As Y. Balasko (1988, p. 11) states: "... the idea that discontinuity is in itself harmful, synonymous of catastrophies (sic), is widespread ... We shall content ourselves with the idea that, from an economic point of view, a continuous evolution path is superior to any discontinuous one."

We want to emphasize, however, that our results only provide the *opportunity* to piecewise continuously adjust equilibrium values during an evolution of the economy. We do not claim to model the real functioning of actual economies.

Another field of fruitful application of our results is that of time consuming equilibria adjustment processes. These generally face a moving target. From our results we can ensure that there actually is a moving target which at least piecewise tracks a continuous path in the space of equilibrium variables. Moreover, we present a standard way to find targets following unavoidable discrete jumps.

In Section 19.4 we will give an integrated analysis of continuously tuning equilibrium variables *and* economic state parameters (cf. Lehmann-Waffenschmidt 2005). We can show that the aim of such a permanently continuous 'double' tuning, or 'fine tuning' in the language of political practitioners, can actually be achieved – if the economic agency is willing to backtrack partly in the control parameter path, at the price of giving the impression of being somewhat undecided – and if the behavior of the economic agents, which is steered by the policy control parameters, does not change over time.

The economic pros and cons for such a double fine tuning of economic solution variables and state parameters have been forwarded in the debate on piecemeal (or gradual) versus bang-bang (or shock, cold turkey) tax reform and in the debate on macroeconomic optimal policy design. To summarize the pro arguments, the shock method roughly speaking produces two types of costs: administrative costs of enactment because of institutional impediments, and social as well as political costs for politicians in a democracy and the society as a whole. Again, our analysis only ensures the opportunity for some exogenous policy agency to achieve a double fine tuning, but neither endogenizes the agency, nor pretends to model real processes in actual evolving economies.

17

Applications of the Analytical Results From Part II in the Economist's Laboratory

In Chapter 14 we have discussed two approaches from the literature which are closely related to ours and we have pointed out how our findings differ from theirs. In this Chapter now we are going to present two further lines of application of our results. Actually, they are also atemporal, i.e., so to speak, they are applications in the economist's laboratory.

The first line of application to be presented below leads to a generalization of the well-known path following method for computing equilibria. To be sure, the conventional application of the path following method only works for regular economies. Apparently, our result of the existence of (near-) equilibrium paths naturally lends itself to a generalization of the path following method to non-regular economies. In Section 17.1 we set out this argument and discuss its gains and losses with respect to the conventional "regular application."

Our second application in Section 17.2 below pertains to the field of comparative statics. Comparative static analysis notoriously suffers from the indeterminateness of the equilibrium set. In fact, how should comparative statics work reasonably when it is not clear which equilibria shall be compared at all? However, multiplicity of equilibria is an inherent trait of the most familiar equilibrium models, and particularly also of those introduced in Part I of this study. Adopting an intuitive wider understanding of the notion of comparative statics which we call 'genetic comparative statics' we show that the findings of our study are well suited for rescuing comparative statics from this seeming dead-end.

17.1 Extending the Path-Following Method to the Computation of Equilibria of Non-Regular Economies

When looking at the figures in the previous chapters showing equilibrium paths the reader may well have the impression that it should be possible to solve the problem of computing an equilibrium by simply following an equilibrium path. More precisely: given some economy with an unknown equilibrium set one has first to choose some well-understood economy, with a unique equilibrium for instance, and a simple connection economy from there to the economy in question. Now, following any equilibrium path of this connection economy one finally ends in an equilibrium of the unknown equilibrium set.

In fact, this method appears to be quite appealing when the economy in question is regular so that one also may choose a *regular* connection economy (see Section 14.2). Actually, this is exactly the approach adopted in the literature dealing with *path following methods (homotopy continuation methods)* for *computing equilibria* (see e.g. Leininger (1978), Allgower and Georg (1980), Scarf (1982), and Mas-Colell (1985, pp. 207-214, pp. 242-243, for surveys). Let us accordingly call it the *regular path following method*. As Leininger (1978, Section 5.4) shows the regular path following method is equivalent to the celebrated global Newton method by Smale (1976). It is the ultimate goal of this approach to provide *explicit* algorithms for following equilibrium paths. Naturally, this is achieved by implicit systems of differential equations (the reader is referred to the cited literature, particularly to Leininger (1978, Section 5.3) and Mas-Colell (1985, pp. 207-214). In Leininger (1978, Chapter VI) one finds a numerical path following algorithm which is well suitable for practical implementation.

After this brief report on the regular path following method we start now our discussion with the obvious remark that it can only be applied to theories where an elaborated regular theory is available. Given a regular theory, however, there still remains the apparent shortcoming of this approach that it only works for regular economies. Moreover, employing a well understood initial economy with a *unique equilibrium* one can only reach equilibria with index +1, see Mas-Colell, 1985, pp. 209-210. Furthermore, there is apparently no reasonable way to generally extend this method to the computation of equilibria of critical economies. Let us show why.

It might be argued that the regular path following method could reasonably be extended to computing of equilibria of non-regular economies. In fact, one might think of the following two natural suggestions to achieve this. To be sure, both methods ground on the principle that approximating a non-regular economy by an evolution of regular economies also admits an approximation of the unknown equilibrium set of the non-regular economy under consideration. Putting it informally, this is due to fact that all equilibria of the given non-regular economy must lie in the closure of the equilibrium sets of the approximating economies.

In order to fix ideas let us confine ourselves to the Walrasian exchange framework. The first suggestion resorts to the reported construction of a well-understood connection economy, the second one makes essential use of Mas–Colell's result on the open–ensity property of the subspace of regular evolutions of economies (see Section 14.2). Unfortunately, it will turn out that none of these suggestions provides a viable extension of the regular path following method.

Let us start with the *first suggestion*. Let any critical Walrasian exchange economy ζ_c be given with an unknown equilibrium set. Now, following Mas–Colell's construction (see Section 14.2) one may choose the well behaved economy

$$\zeta_q : \mathring{\Delta}^{n-1} \longrightarrow \mathbb{R}^n$$
$$p \mapsto \left(\frac{1}{p \cdot q}\right) q - p$$

for some suitable $q \in \mathring{\Delta}^{n-1}$ and any regular economy ζ_a near ζ_c such that the convex connection economy from ζ_q to ζ_a is regular. Starting from the unique equilibrium q of ζ_q, and following an equilibrium price path one finally clearly arrives at some equilibrium of ζ_a. Approximating ζ_c with a sequence of regular economies (ζ_a^k) one could hope to achieve an approximation of an equilibrium of the given economy ζ_c.

Let us now check this method more carefully. Obviously it requires that at least all economies of a whole tail of the sequence (ζ_a^k) lie in the same component of the regular subspace. This component furthermore has to satisfy the condition that its closure contains the critical economy ζ_c. But, even when one has found a regular economy ζ_a^1 from such a component it can be a fairly tricky problem to find a sequence (ζ_a^k) with the desired properties. For instance, there might be a situation like that of Figure 17.1 where the convex connection economy between any ζ_a^k and ζ_c crosses infinitely many regular components. This, clearly, makes the whole suggestion useless. The following simple example in one dimension actually produces such a situation (see Figure 17.2 below).

These problems likewise *also* apply to the *second suggestion*: as before, one starts with a regular economy ζ_q with the unique equilibrium q. Now one considers the convex connection economy $\overline{\zeta_q \zeta_c}$ between ζ_q and ζ_c. Clearly, it is non-regular. But from Mas-Colell's open-density result (see Section 14.2) one may choose some regular evolution $(\zeta_s)_{s \in [0,1]}$ arbitrarily close to $\overline{\zeta_q \zeta_c}$. Apparently, $(\zeta_s)_{s \in [0,1]}$ can be chosen so that the initial state economy ζ_0,

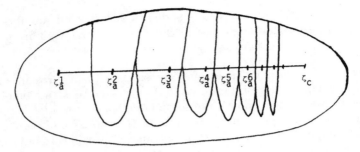

Fig. 17.1: Convex Connection Economy Crossing Infinitely Regular Components

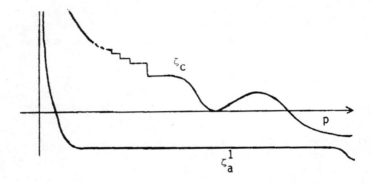

Fig. 17.2: Function Generating a Convex Connection Economy Crossing Infinitely Regular Components

moreover, has a unique equilibrium close to q. Nevertheless, it is obvious that one has the same problems with the crucial terminal state ζ_1 as before with the approximating sequence (ζ_a^k). Thus, also the second suggestion does not work satisfactorily.

Now, this is where our results from Part II come in: *Just choose some simple connection evolution (see Section 11.3) from any well-understood initial economy to the economy in question, and follow an equilibrium path.*

Actually, our method is much more flexible since all mentioned limitations evidently do not anymore play a role for it. To be sure, however, this is achieved at the price of lacking explicit numerical path following algorithms.

17.2 Generalizing Comparative Statics to Economies With Multiple Equilibria

Comparative statics notoriously suffers from indeterminateness: comparing two states of a model with multiple equilibria the troubling question arises which equilibrium of the first state has to be compared with which equilibrium of the second state. As Arrow and Hahn put it (1971, p. 245): "[The problem of] giving unambiguous predictions of how the equilibrium of an economy will be affected by a given parameter change ... must be intimately related to that of the *uniqueness* of an equilibrium, and it is pretty clear that we shall not expect to get very far without stipulating one or the other of the conditions that ensure such uniqueness. Even so, the kind of parameter changes for which predictions become possible is pretty limited."

Fourteen years later Kehoe (1985a, Introduction) even sharpens this negative conclusion: "Conditions that guarantee the uniqueness of equilibrium in models of economic competition are crucial to applications of these models in exercises of comparative statics. The fundamental hypothesis underlying this type of analysis is that the state of the economic system can be completely specified by the solution to a mathematical model, which is the equilibrium of the system. If, for a given vector of parameters, there is more than one solution to the model, then the comparative statics method breaks down. Lacking conditions that guarantee uniqueness, we must resort to considerations of historical conditions and dynamic stability, which greatly complicate the analysis."

This is what literature tells us. But is this really the only possible conclusion from the multiplicity of equilibria? Does it definitely mean the deadlock of comparative statics when equilibrium sets are non-unique? It does not. What we want to say is that this negative view comes from a much to narrow understanding of the notion of comparative statics. In our eyes a broader understanding would be more adequate. Not the comparison of two unique equilibria at distinct states of the economy is at the very heart of comparative statics, but, more generally *the opportunity to uninterruptedly pursue the evolution of equilibria while the economy continually evolves from one state to another*. From this *"continuous"*, or *"genetic comparative statics"*, as one might call it, we learn *how* equilibria of the terminal state have originated from equilibria of the initial state. And this is what our result of the existence of (near-)equilibrium paths ensures for each of our basic models. In addition to that Chapter 8 shows that – at least for the Walrasian equilibrium framework – this is precisely the only equilibrium structure property which generally holds for evolutions.

The Method of Kinetic Analysis of Evolving Economies in Historical Time

Tempora mutatur, et nos mutamur in illis.

from medieval Christian Latin

We have still to consider the economical condition of mankind as liable to change, and indeed as at all time undergoing progressive changes. We have to consider what these changes are, what are their laws, and what their ultimate tendencies; thereby adding a theory of motion to our theory of equilibrium.

J.S. Mill

In this Chapter we will leave the purely analytical and atemporal viewpoint mostly adopted in the previous chapters turning to a temporal conceptualization which in the literature has been labelled "kinetics", or "kinematics". Like the terms *static* and *dynamic,* also the term *kinetic (kinematic)* has a long tradition in physical sciences. There it means the *"description of the motion of objects without considering the forces that cause or result from the motions"* (Encyclopaedia Britannica, Micropaedia, 15th ed. 1985). In the economic literature the term 'kinetics' for this kind of analysis first appears in the early twentieth century (F. H. Giddings, 1911; F. Oppenheimer, 1916/19/23) with a status somewhere between comparative statics and dynamics (see Ott (1970), pp. 16–20, for a historical survey).

The purpose of this Chapter is to lay the methical foundations for our subsequent analysis of economic systems evolving in historical time in Chapter 19, particularly in the sections 19.2-19.4. Thus we first will elucidate the relationship of kinetics to comparative statics and comparative dynamics. Due to the traditional narrow understanding of comparative statics requiring a unique equilibrium only very few and limited comparative static results which may be viewed as kinetic ones can be found in the literature (e.g. Arrow/Hahn (1971), Chapter ten, Theorem 5; Quirk/Saposnik (1968), Section 6.4, Theorem 5). This encourages us here to maintain the term 'kinetic' for characterizing our approach which is essentially different from the traditional comparative static approach. In this Chapter we will first give a full characterization of the kinetic method in the specific context and purpose of our study.

Nowadays, there is no more disagreement among economists what has to be understood under a *dynamic system* or *model*. Actually, a dynamic model must provide an explicit explanation of *how the states* of the economy system under consideration *evolve over time* (cf R. Frisch's influential article from 1935). More detailed, a dynamic model must explain for each state *how* it developed from the previous state(s), and *how* the future state(s) will develop from the present one (see for instance the "classical" contributions on this issue by Smale (1981), Sonnenschein (1986), or Batten et al. (1987)). Accordingly, the traditional mathematical devices for formal treatment are difference equations, which lead to a discrete 'period analysis', or differential or integral equations, which in turn lead to a continuous 'rate analysis'.

The new branch of *evolutionary economics* widely dispenses with the idea of describing the evolving economic system by systems of differential, or difference equations. It is characterizing for the evolutionary approach that it gives an 'open' modelling of the evolving economic system, that means that it does not predetermine the course of the described system completely as systems of differential equations do it. Actually, when considering evolutions as course evolutions we also adopt the *open loop modelling* view. Moreover, our approach has in common with the evolutionary approach that it does not view equilibria as describing the actual states of real economic systems. In fact, for both approaches equilibria play solely the role of crucial points of reference. Nevertheless, there is a great difference between the two approaches as regards the aim of investigation. We will come back to this point below when we will discuss the differences between the dynamic and the kinetic approach.

However, there is unanimity under the profession that only little progress has been made so far towards a realistic comprehensive dynamic theory. (Moreover, there is even a widespread pessimism whether such a theory can be achieved at all.) Anyhow, the consequence of this is that the economist is usually left to a *comparative static analysis* in some static framework when she wishes to study a changing economic system. Apart from the limited comparative static results in the literature which have a kinetic character and which we have addressed before a comparative static analysis in the traditional sense usually either gives a result on the *local tendency* of the endogenous variables when the exogenous parameters are infinitesimally varied, or it *compares* the values of the endogenous variables before and after a discrete change of the exogenous parameters.

After this brief review of the comparative static method let us recall the standard criticism on the 'discrete comparative statics' mentioned last. Contrasting two distinct states of an economic system one does not take any care of *how,* or even of *whether at all,* the second state may evolve from the first one. As F. Fisher puts it (1983, pp. 15-16): "Comparative static analysis will show us how the equilibrium corresponding to a particular set of circumstances changes when a particular parameter shifts. However, the displaced equilibrium will not be that to which the disturbed system converges (assuming stability). Rather, the very process of convergence, of adjustment to the displacement, will itself further change the equilibrium. Whether or not the ultimate equilibrium will be close to the one predicted by comparative statics, or even whether the ultimate effects of the displacement will be in the predicted direction is not a question that lends itself to a general answer. The answer depends on the effects of the parameter shift on the adjustment path of the system, on comparative dynamics rather than on comparative statics. Unfortunately, a satisfactory analysis of comparative dynamics lies in the future."

Summing up, both lines of comparative statics – the infinitesimal and discrete comparative statics – have the deficiency in common that they provide no information at all about the *two paths* between (1) the two states of exogenous parameters, and between (2) the two states of endogenous variables, respectively. In particular, conventional comparative statics has a very limited scope when there are *multiple equilibria* as for instance Kehoe (1985a) emphasizes (cf. Section 17.2).

Now, let us come back to the kinetic approach in economics. Unlike the conventional two lines of comparative statics the kinetic analysis is neither confined to *comparing distinct states* nor to giving just *qualitative statements on the effects* on the endogenous variables caused by infinitesimally small variations of the exogenous parameters. Instead, a genuine kinetic analysis: like ours *continually* studies the effects of unrestrictedly exogenously changing state parameters, or even state functions, on the endogenous variables. More specifically, it is the *general aim of kinetic analysis to examine the evolution of the dependently changing endogenous (solution) variables resulting from an unrestricted (, but continuous) exogenous evolution of state parameters, or state functions, for general structure and regularity properties* (cf. e.g. Ott (1970), pp. 16-20).

So far, we only have pointed out the differences between the kinetic and the comparative static approach. Hence it still remains to clarify the particularly interesting *relationship between the kinetic and the dynamic approach.* In contrast to dynamic modelling the kinetic approach neither pretends, nor even attempts, to explain the evolution of state determining parameters, or state functions. What kinetics is purposed for is the investigation of the induced evolution of the dependent endogenous equilibrium variables for general structural properties. Using the terminology of Faber/Proops (1998) the evolution of state determining parameters, or state functions, is considered 'genotypic', and thus exogenous and unexplicable. On the other hand, the merely 'phenotypic' evolution of the dependent endogenous equilibrium variables is well amenable to economic analysis since it is completely determined by the underlying genotypic evolution of states. Speaking formally, the kinetic approach treats both evolutions as "quasi" functions of elapsing time and does not investigate into the causal intertemporal relationships between successive states. Thus, both addressed evolutions can be viewed as merely 'phenomenological' recordings. Actually, there is made widespread use of such recordings in daily life. Just think for example of seismograms, electro-cardio- and electro-encephalograms, time profiles of aggregate economic magnitudes for business cycle studies, or asset charts. As an illustration Figure 18.1 shows the chart of prices of the Infineon Technologies share at the stock exchange of Frankfurt from Januar to December 2003.

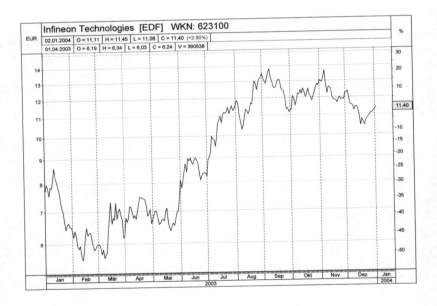

Fig. 18.1: Chart of the Infineon Technologies Share

However, the numerous attempts of the so-called 'chart-analysts' to develop prediction rules for future stock prices from characteristic chart patterns in the past bring it to light how ambitious it is to strive for a comprehensive dynamic theory with true predictive power. In last consequence this amounts to the archetypal human desire for being omniscient, or even omnipotent, which, as we are told by the bible, was the cause for the expulsion of mankind from Paradise.

Let us summarize: in the spirit of our epigraph "$\Pi\acute{\alpha}\nu\tau\alpha\ \overset{c}{\rho\tilde{\epsilon}\tilde{\iota}}$." ("everything is fluctuating") kinetic analysis generally deals with the evolution over time of some modelled economic system. In contrast to dynamic analysis, however, kinetic analysis makes no theoretical prespecifications of intertemporal influences. This means, no intertemporal restrictions are imposed on the evolution of states of the economic system which is thus just taken as exogenously generated in a black box. To repeat it, kinetic analysis - faute de mieux - is not concerned with the causal explanation of the intertemporal forces which determine the evolution of states. What kinetic analysis is concerned with is the search for *general regularity properties* of the *evolution of the dependently changing endogenous (solution) variables. Thus, one could also say that the goal of kinetics is not to investigate 'the law of the evolution of the economic states', but rather 'the law of evolution of the endogenous dependent (solution) variables'.*

The following Figure 18.2 surveys the methodological position of the kinetic approach. Start with choosing some basic model. The path e^I drawn in

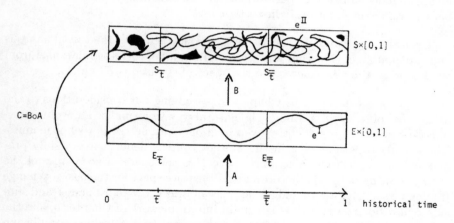

Fig. 18.2: Methodological Position of the Kinetic Approach[1]

---

[1] The reader may have noted that e^I and e^{II} in Figure 18.2 are not quite correctly pictured in so far as the mapping B does not assign identical momentary equilibrium sets to identical states of the economy, like $E_{\bar{t}}$ and $E_{\bar{\bar{t}}}$ for instance, as it should do it correctly, i.e. in Figure 18.2 the t-equilibrium-slices $s_{\bar{t}}$ and $s_{\bar{\bar{t}}}$ should have an identical geometrical shape. However, we have accepted this inconsistency in Figure 18.2 for the sake of a greater easiness in drawing the evolution e^{II}.

the product space $E \times [0, 1]$ of the space of economies E and the time interval $[0,1]$ symbolizes the evolution of states of the modelled economic system. Of course, e^I *projects one-to-one* on the time interval ('one state at any moment of time'). The path e^{II} depicts the evolution of the endogenous solution variables in the product space $S \times [0, 1]$ of the endogenous variable space S and the time interval $[0,1]$. Due to the indeterminateness of the underlying basic model the evolution e^{II} in general is not a path projecting one-to-one on the time interval, but a complex subset, or correspondence, which projects *onto* the time interval ('at least one solution at any moment of time'). The arrows A and C symbolize the observed and thus purely *phenomenological, non-causal* relation between elapsing time and the recorded evolutions e^I and e^{II}, respectively. In contrast, the arrow B denotes the deterministic relation how the model associates the corresponding endogenous solution variables with any state of the economic system. (Thus, $C = B \circ A$.)

Now, kinetic analysis takes the *'genotypic' evolutions* e^I as exogenously given, and examines the *'phenotypic' evolutions* e^{II} for general regularity, or say structure, properties. This is done in the hope for achieving more transparency and better predictability of what may happen at the phenotypic level at all. Generally spoken, it is the ultimate purpose of kinetic analysis to explore the opportunities for precautionary intervention by some governing institution in favor of a more satisfactory performance of the economic system. We will come back to this issue extensively in Section 19.4.

Evolutionary economics, in contrast, is concerned with the causal analysis and explanation of the evolution of economies e^I, i.e. with the relation A from the historical time parameter t to the space of evolutions $E \times [0, 1]$.

Let us now conclude this Chapter by specifying our considerations to the analytical context of our study. The natural way to formalize an evolution e^I of economic states is evidently given by a one-parametrization of economies. The evolution parameter s in the present context denotes irreversibly progressing historical time. The state space E is the space of economies of the chosen basic model, and S accordingly, is the price space for the basic exchange frameworks of Chapters 4 and 5, the space of prices, tax redisbursals, and subsidy realization rates for the tax equilibrium framework of Chapter 6, and the space of quantity rationing bounds for the quantity constrained equilibrium framework of Chapter 7, respectively. In our context of general equilibrium the normative terms in the last two sentences of the preceding paragraph may most naturally be specified as continual satisfaction of all individual plans in a most 'frictionless' way. What we precisely mean by 'frictionless' will become clear in Chapter 19 below. For the moment let us content ourselves with the basic idea that frictions are mathematically reflected by discontinuities and, to use Balasko's words, that "discontinuity is in itself harmful, synonymous of catastrophies". This clearly makes "a continuous evolution path superior to any discontinuous one" (Balasko, 1988, p. 70).

Having reached this point we still have to take up an argument which has been prominently advocated by J. Hicks (1965, 1985). Putting it short it is along the following lines: describing an economy in process by a sequence of static models can never be adequate since at any date all economic relationships are solely between current variables. As Hicks puts it:

"I began with the assertion that there is equilibrium when all 'individuals' are choosing the quantities, to produce and to consume, which they prefer. To a conception of equilibrium that is of this type we must hold fast. But how can we make these quantities dependent (in a dynamic economy) upon current parameter the equilibrium values of time t upon the parameters of time $t-$ and upon those only? The question did not arise in a static model, since the parameters, on which the equilibrium depended, were at all dates the same. Here they are not the same. If (say) population is increasing, an 'equilibrium' that is based upon present population, paying no attention to the increase of population, will not even be a *transitory equilibrium;* there will be no reason why the 'individuals' should leave the population movement out of account in their investment decisions; there will be no reason why there should be even a 'tendency' in the direction of an equilibrium that is solely based upon present population. Similarly for other variables. The static equilibrium, entirely based upon current parameters, is in strictness irrelevant to the dynamic process. (...)

It is (...) that the equilibrium of time t could be taken to be determined by *current* parameters only: or, as we may put it now that we are using a sequential framework, that the equilibrium of the single period may be treated as *self-contained.* In a fully static theory this is a perfectly harmless assumption. Nothing has to be said, in statics, about the obvious point that production takes time, so that it must be oriented, not towards the present, but towards the future; for if present and future are identical we can substitute one for the other without making any difference. We can take a demand curve (for instance) which reflects current wants, and set against it a supply curve that refers to current supply; for the same demand curve will still be 'there' when the process of production is completed; we do not have to bother about the fact that they refer to different *times*. But in dynamics these things do matter; it is of the essence of the dynamic problem that present and future are *not identical*.

Proper dynamic theory, even at its single-period stage, must take account of the fact that many activities that go on within the period are oriented outside the period; so that what goes on, even within the period, is not only a matter of tastes and resources, but also of plans and expectations. In statics there is no planning."

So far the argument by Hicks on what he calls the 'static method in dynamic theory'.

Now, what are the consequences of this for our kinetic method? Reading Hicks' line of argumentation for the first time might give the impression that only his position can be a valid and sound base for a temporal analysis. However, reading once more carefully what Hicks writes changes the picture substantially. Hicks' argument essentially points to the main problem of theoretical economics, namely which variables are treated as endogenous when building a model, and which ones are left outside in the exogenous 'black box'. Now, considering our evolutions introduced in Part I as truly temporal evolutions, i.e. as describing evolutions in historical time, we are not liable to Hicks' reproach for the following reason. Taking any one of our basic models and viewing it as a snap shot from an ongoing economy at a certain date, each functional economic relationship used there can naturally be viewed as already embodying all intertemporal interrelations caused by the agents' experiences from the past and by their expectations for the future. In other words, the dated functional relationships of the snap shot are considered as black box as regards the intertemporal influences. What is explicitly and endogenously related by the 'snap shot functional relationships' are solely contemporaneous variables (prices, quantity constraints, tax rates, subsidy rates) which are at the current disposal of the economic agents.

Summing up our argument, the performance of the economy in the course of time is represented by a succession – be it discrete or continuous – of 'black box snap shot models'. The characterization 'black box' accounts for the lacking explicit causal intertemporal interconnectedness. Consequently, at any date only current variables are left to the agents' current disposal, and it is this set of current variables from which the chosen model selects the variables which are related by the dated functional relationships describing the agents' economic behaviour.

Before we can proceed to the next chapter we still have to settle a peculiar issue concerning the term 'temporary'. Though it is tempting for us to use the term 'temporary' for our snap shot economies, i.e. the state economies, of an evolution, we, nevertheless, will not do so in our study since it easily could lead to misunderstandings. Actually, already there are at least two modes of usage of the term 'temporary' which are well established in the literature.

The first one is that used by Hicks in his famous 'flexprice method' (1946, Ch. IX, 1965, Ch. VI), and the second one that used by the French-Belgian fixprice quantity-rationing school initiated by Malinvaud, Drèze, and others (see e.g. Grandmont (1982) for a survey; cf. also Chapter 7 of this study).

However, our approach does neither attempt to explain the formation of actual prices in an economy, as Hicks does, nor does it always use fixed or rigid prices as the quantity-rationing school does. To repeat it, our primary concern in this study is with the analysis of possible regularities which generally hold in the equilibrium sets of evolutions of economies. In other words, we do not investigate whether, or how, equilibria are attained in actual economies; in particular, we even do not claim that observed prices in actual economies are equilibrium prices in the defined sense. Thus our position is much less ambitious than the usual position of equilibrium theorists. Throughout the whole study we just view equilibria as crucial points of reference – from the economic and drom the analytical viewpoint – which, nevertheless, do not necessarily reflect real phenomena.

After these clarifications on the relationships of our approach to well-established branches of economic theory we come back to the main line of our exposition. In the remaining two chapters of our study we will employ our previous conceptual insights and analytical results for analyzing ongoing economies in the kinetic mode. Not surprisingly, also with our kinetic approach we have the well-known choice between the two familiar ways of modelling elapsing time. This means, either to divide up time into a sequence of finite periods, which means a *discrete analysis*, or to let periods have infinitesimal length, which leads to the limit case of *continuous analysis*. Following this categorization Section 19.1 will be devoted to the discrete and the Sections 19.2–19.4 to the continuous case.

Evolving Economies in Historical Time

> Economic life is extraordinarily continuous, characterized by a getting and spending which does not even cease at night.
>
> Richard M. Goodwin

> As the variables that are usually considered and observed by the economist are the outcome of a great number of decisions taken by different operators at different points of time, it seems natural to treat economic phenomena as if they were continuous. ... A further difficulty of discrete analysis is that usually there is no obvious time interval that can serve as a "natural" unit.
>
> Giancarlo Gandolfo,
> Pietro C. Padoan

When modelling ongoing economies in historical time theoretical economists usually resort to the discrete period approach referred to in Chapter 19.1. Nevertheless, the subdivision of the time axis into discrete periods raises several serious questions:

(1) How long are the periods?

(2) How can the period length be the same for all markets?

(3) How sensitively does the whole analysis depend on the chosen period length?

(4) How sensitively does the whole analysis depend on the choice of the position of the origin of the chain of periods?

Obviously, these problems are in no way marginal ones. Though it is clear that a period naturally means a planning period the controversy about the length of the periods is pervasive in economics. So let us have a closer look at these problems. Particularly, it is hard to conceive the same planning period for all markets. Just think, for example, of the markets for satellites or nuclear power stations on the one side, and for everyday's necessities on the other side. Finally, it is well known from descriptive statistics that changing the width of class intervals, or shifting the class intervals on the axis in general has serious effects on the analytical results.

A way out of these shortcomings of this 'brick period' approach, as we will call it, has been provided by the family of growth models which are formalized by systems of differential equations. In contrast to 'brick period' growth models which are formalized by systems of difference equations these models basically consider the evolution of economic states as a *continuous (and even differentiable) flow* of the relevant economic magnitudes. Formally spoken, one lets converge the length of the 'brick' periods to zero, thus obtaining in the limit a continuum of infinitesimal, or 'degenerate', periods.

In this Chapter we are going to model and to analyze economies evolving in continuous historical time. In Section 19.2 we will adopt the *flow view* we have addressed above. However, we want to emphasize that it is only the method of modelling the evolving economy by means of continuous flows which our approach has in common with the approach of differential growth models. In fact, we do not use differential equations; and particularly *we do not predetermine* the evolution of economic states as it is done when a system of differential equations together with the initial conditions is specified. Consequently, there is no growth tendency nor any cycles inherent in our modelling of an evolving economy. Instead, our conceptualization is far more flexible in that it *admits at any state* of the evolving economy *any ensuing state* - as long as it does not violate the continuity assumption. In the literature on economic evolution theory one also finds the term 'open' for our type of modelling evolving economies, and 'closed' for the predetermining type.

This Chapter is organized as follows. Sections 19.2.1 and 19.2.2 present two alternative methods of modelling an ongoing economy over continuous time. Strictly speaking, we are dealing with ongoing economies which are 'left on themselves', i.e. with course evolutions in our terminology from Chapter 12. In Section 19.2 we introduce the 'flow modelling' which in some sense is in the spirit of differential growth models. As we will see, eight of our nine basic models from Part I well suit as basic set-ups for this approach. The exceptional role of the quantity constrained basic micromodel from Section 7.1 is due to the fact that it is the only basic model whose functions represent individual, i.e. non-aggregate behavior. In Section 19.2.2 we propose a new way of modelling ongoing economies over continuous time which is appealing both from the viewpoint of economic intuition and of formal tractability. The basic idea is to completely give up the idea of successive (brick or infinitesimal) time periods, and to use a continuously 'sliding period' instead. For evident reasons only the two basic models from the exchange framework without Walras' law and homogeneity from Chapter 5 suit as basic set-ups for this approach. Thus, we have further justification for relaxing the traditional assumptions of Walras' law and homogeneity.

In a nutshell our conclusion will be that for any ongoing economy in one of the admissible basic settings it is possible to adjust (near-) equilibrium values *piecewise continuously* during elapsing time. The reader should well note, however, that we only establish the opportunity of doing so, but do not strive for a theory whether, or how, (near-) equilibrium values from the evolution of equilibria are actually selected when the economic system evolves. What are the advantages of the opportunity of adjusting equilibrium values (piecewise) continuously over time? One advantage is most obvious: following a continuous path makes it fairly easy to find new equilibria for new economic states. Further advantages become plain if one thinks of the effects of abrupt discontinuous changes of equilibrium variables. Surely, even then the aim of permanent and simultaneous balancing of all markets, the 'homeostasis', is attained. But, discontinuous changes of equilibrium variables in general also cause discontinuous changes of the economic statuses of the agents. Due to the widespread attitude of risk aversion and conservatism in economic affairs discontinuous changes of economic conditions as well as merely the expectation of discontinuous changes are highly unwelcome in reality (see also Balasko, 1988, p. 70).

In Section 19.3 we will provide a further application of our findings. Probably Rosenstein-Rodan (1930) was the first to make the problem precise that any equilibrium adjustment process which is not infinitely fast in general faces a . In his contribution F. Fisher (1983) revisits this problem. In Section 19.3 we demonstrate that the setting provided in Section 19.2.1 together with our earlier analytical results allows for a thorough general analysis of this problem.

So far we have dealt with ongoing economies only which are left to themselves, i.e. with *course evolutions* in our terminology. At this point the natural question arises whether our theoretical setting also provides the opportunity for some external political economic agency, say a governmental authority, to beneficially intervene in the evolution of economic states. On the one hand one either may think of interventions in a *course evolution* from time to time, or, on the other hand, of a complete control of a whole connection evolution. Obviously, this requires explicit control parameters. Fortunately, some of our basic models contain parameters which, at least in principle, can be controlled by some external agency. These are the two basic models with production, taxes, and subsidies from Chapter 6, and the quantity constrained multi-sectoral model from Section 7.2. In the first case it are the tax and subsidy schemes which are controllable, in the second case it is the price/wage system. For these set-ups we show in Section 19.4 that for any *course* evolution there is the opportunity for some external agency to adjust (near-) equilibrium values without any discontinuous leaps by implementing finitely many continuous repetitions, or backtrackings in the evolution of economic states. Moreover, the same is true for any prescribed *connection evolution* from some undesired state of the economy to a desired one (cf. Lehmann-Waffenschmidt (2005)). We want to emphasize, however, that we do not model the controlling agency endogenously, but merely study the opportunities open to such an agency. In this sense we also do not provide a theory about which (near-) equilibrium path in the generated evolution of equilibria actually will be selected and which backtrackings in the evolution of economic states are chosen. The reader should note, however, that a connection evolution controlled in this sense comes close to a truly dynamic model.

We have already pointed out in the Introduction to Part III the economic advantages of a permanently continuous, or piecemeal, gradual, frictionless *adjustment of equilibria* when the economy evolves. The advantages of controlling the whole evolution of the economy such that the control parameters as well as the economic states all the time change continuously have been thoroughly discussed in the literature on piecemeal tax reform in the seventies (Feldstein (1976), Hettich (1979)) and also in a branch of macroeconomic optimal policy design (see Gandolfo and others, 1984, 1988). Summing up the tax reform debate there are two categories of reasons speaking against radical alterations of tax schemes ('de novo design', bang-bang, cold therapy, shock therapy policy): first, there are administrative costs of implementation, and second, there are political constraints. The arguments by the macroeconomic policy design economists which are relevant for us amount to the criticism on the unsolved, and perhaps even unsolvable, problem of choosing an appropriate length of the periods which has been mentioned by us at the outset of this chapter and to the continuity-by-aggregation-argument expressed in the second epigraph of this Chapter.

Our results lead to the following conclusions. A controlling governmental authority in general is forced to partially and repeatedly backtrack in the path traced by the control parameters during the course of time. Or in other words, when the controlling agency has the aim to avoid radical changes of the state of the economy as well as of the equilibrium values it, in general, has to backtrack partially and consequently to accept that it appears to be inconsequent in its actions.

19.1 Evolving Economies in Discrete Historical Time

> Any process of change can be exhibited, if we choose, as a sequence. The process is divided into steps, or stages, which can be taken separately, and analyzed separately ... Business men think in time periods, and it is in terms of time periods that they do their accounts.
>
> J. Hicks

The common economic understanding of any of our basic models from Part I is that it represents the average performance of an economic system during a single period of time. Thinking of time as subdivided into periods of finite length this view readily suggests to model the performance of the economic system over time in the kinetic mode by a sequence $(E_t)_{t=1,2,...}$ of model specifications, or states. The economic justification of this has been given in the last part of Chapter 18 where we defended our approach against the Hicksian argumentation. We will also call $(E_t)_{t=1,2,...}$ a *discrete evolution over time* based on the basic model E.

At the first glance this procedure seems to be in complete analogy to the transition from a single snap shot to a cinematographic, or say a *stroboscopic,* time and motion study where a movement process is not scanned continuously, but recorded at discrete points in time producing a series of individual images. Nevertheless, the reader should be well aware that there is a significant difference between the succession of economic states as we have described it and a cinematographic series of shots of a movement process. This concerns the meaning of the time interval between two successive events, i.e. market dates, or shots.

While the time interval between two successive single shots simply measures physical time, the time period between two successive market dates in addition has a social meaning. It represents the agents' planning period for their economic activities which become materialized at the second market date.

Before we are going to exploit our previous results for the presented discrete conceptualization of evolutions over historical time we would like to emphasize that all considerations in this Chapter equally apply to each of our basic models from Part I. This will be no longer the case when we will study the continuous conceptualization of evolutions over time in Chapter 19.

Now, what are the merits of our results from Part II for the *discrete kinetic conceptualization* of an ongoing economy in historical time?

To the reader who has read the applications of our formal results in Chapter 17 the answer will be obvious. In fact, the following two general questions are of paramount interest to the economist when modelling an ongoing economy in the described discrete way: *firstly*, how to compute equilibrium vectors for any state economy of a discrete evolution – whether it is regular or non-regular. And *secondly*, how to compare equilibrium vectors of two different state economies, whether they have unique or multiple equilibria.

These are our achievements:

To generalize the path following method to non-regular economies in Section 17.1 we have developed a tool to cope with the first question.

To tackle the second question one needs a method which is also capable of managing with multiple equilibria. This has been achieved by our generalized continuous comparative static method which we have developed in Section 17.2.

Thus, the applications of our previous conceptualization and results to economic systems evolving in discrete historical time are straightforward. The picture will be more complex, however, when we turn to a continuous understanding of elapsing historical time in the following Chapter 19. This step, nevertheless, will be rewarding from the economic viewpoint since there are severe problems with the discrete approach as we will see immediately.

19.2 Alternative Models of Evolving Economies in Continuous Historical Time

19.2.1 Flow Commodity Models of Piecewise Continuously Balanced Evolving Economies

In this Section we will outline our first method how to design kinetic models of evolving economies in *continuous historical time*. As mentioned in the Introduction to this Chapter the basic idea has been adapted from a certain constructing principle of growth models. Accordingly, the economic process is described by continuously fluctuating flows of commodities. These flows are formally represented by continuously fluctuating *cross-section magnitudes*. Thus, at any single moment the *size of a commodity flow measured in units of quantity* clearly is zero. On the other hand, the *momentary geometrical cross-section size of a flow* is a well-defined nonnegative real number. As an illustration just think of a faucet with variable opening size where the water flows out with constant velocity. The momentary cross-section size of the flow of water at date t_0 is the area of the opening size of the faucet at date t_0.

Following the principle that a continuous modellation over time is economically meaningful for aggregate economic activities, this general approach makes sense *for all of our basic models* from Part I, except for the quantity constrained micromodel from Section 7.1. This is due to the fact that the basic models of the first group exclusively use *aggregate economic behavior functions,* whereas the primitives of the quantity constrained micromodel are functions of *individual behavior.*

Going through the first addressed group of basic models we see that any function used there naturally represents a real or a nominal flow of economic quantities: demand, supply, and excess demand of commodities, services, labor assets, tax receipts and redisbursals, and subsidies.

More precisely, the basic idea with our *continuous flow method* is that the independent variables which occur in our basic models, i.e. prices, budgets, tax redisbursals, and subsidy rates, are viewed as *'ultra-short-run',* i.e. *momentary, determinants of momentary economic behavior.* To be more specific, prices and subsidy rates are considered as determining *momentary cross-section* demand, supply, excess demand, and tax payments respectively. This intuitive kinetic conceptualization of evolving economies in continuous time naturally lends itself to a formal representation by evolutions of economies as we have introduced them in Part I. Subsequently we will also speak of *continuous evolutions over time.* The question how to measure quantities and their value over time in this context is straightforward to answer. For example, the demanded quantity and its value over a certain interval $[t_1, t_2]$ are expressed by the natural integrals which are shaded in the following Figure 19.1.

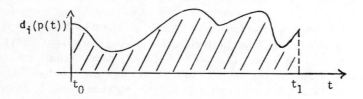

Fig. 19.1: Quantity of Commodity i Demanded in $[0, \bar{t}]$

In other words, we consider any evolution in any basic model from the addressed group as a *continuum of evolving markets,* or of *market events,* over elapsing time. In complete analogy to the static (periodic) interpretation, at any single moment each *cross-section market* will be balanced when the inde-

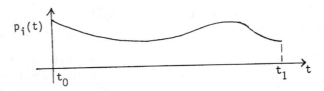

Fig. 19.2: Price of Commodity i

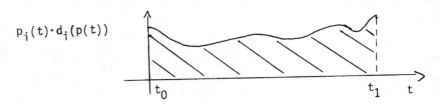

Fig. 19.3: Total Cost of Demanded Quantity of
Commodity i in $[0, \bar{t}]$

pendent momentary variables take equilibrium values. Thus, we have a succession of *momentary* equilibria, which we also may characterize as *ephemeral, or transient, or transitory equilibria*. In Baumol (1970, Introduction) one finds the following simple example with one commodity. The example pictures the history of an economy on a multidimensional graph with time along one of the axes. This is illustrated in Figure 19.4 below which shows the demand for and supply of a particular commodity through time under conditions of perfect competition. The *course of demand through time* is given by the surface $D_1 D_2 D_3 D_4$, which shows how much of the good will be demanded at different prices at each moment of time, and similarly the course of supply over time is given by the surface $S_1 S_2 S_3 S_4$. The course of the equilibrium supply and demand situation, i.e., the *time path* of the price which *continuously* equates supply and demand *(the moving equilibrium)*, is shown by EE'. Now if we take a cross section of the diagram perpendicular to the time axis at time OT, thus in effect considering a very thin slice of the diagram at that time, we have the ordinary static supply and demand curves at time OT, $S_2 S_3$, and $D_3 D_2$ respectively. Thus the static method analyzes a "time slice," a cross section of the economy, thereby eliminating the passage of time from the problem, though, as we shall see, not necessarily eliminating the *influence* of time altogether" (Baumol 1970). Let us add here that the nice non-backtracking feature of the equilibrium path EE' is a consequence of the uniqueness of any momentary, or say 'time-slice', equilibrium. Figure 19.4 shows the 'manger' picturing the economy's history.

We particulary emphasize that in this way we model an open loop evolving economy which is *left to itself*, i.e. there is no restriction for the evolution

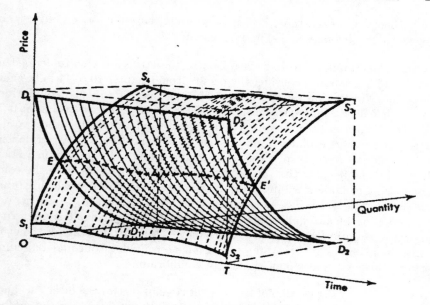

Fig. 19.4: Three Dimensional Graph of a Market Evolution

except for continuity. To put it more stringently, in this Section we do not explicitly include a political economic agency or institution which takes intervening measures of any kind. Speaking in the terms of Section 12.1 we thus are exclusively dealing with course evolutions. Now, what is the significance of our analytical results from Part II for this continuous flow conceptualization of an evolving economy in historical time?

Let us begin with the results from Chapter 13. There we have argued that Theorem 13.1 and Corollary 13.2 extend Mas-Colell's static result from 1977 to the one-parametrized set-up. Now in the light of our kinetic flow interpretation we can even go further and state: these results also give a general characterization of the *degree of indeterminateness* of the exchange framework *on the kinetic level*. There might, however, be objected that our results from Chapter 13 would not really mean a true extension of the static decomposition result in the strict economic sense. In fact, while a static exchange economy with individually rational agents may well be considered as a reasonable one-shot of an economic system, it is by no means clear how to apply the criterion of reasonableness, or rationality, to a certain succession of economic states, be it historical, or not. To this argument we reply the following: to judge how reasonable a particular succession of economic states is, i.e. an evolution of economies, obviously would require well-founded general standards. And it is furthermore obvious that provision of such standards in turn would require a well-founded comprehensive dynamic theory. As long as there is no such dynamic theory, however, there is clearly no sound base for sorting out certain evolutions of economies as being unreasonable.

After this application and discussion of the structure results derived in Chapter 13 let us now reconsider the results from Chapter 10 to 11 in the new light of our kinetic flow interpretation. The existence of a (near-) equilibrium path with backtracking parts means that for any continuous evolution it is possible to *piecewise continuously tune* equilibrium values. In other words, for any continuous evolution one can ensure a 'homeostasis', i.e. permanent perfect balancedness, by piecewise continuous tuning of equilibrium values. The reader should note that arguing this way we completely leave the common stationary understanding of equilibrium as a state of rest which Kornai (1983) calls the "scientific" notion of equilibrium. Instead, to repeat it, our understanding of an equilibrium vector here is that of an ultra-short run, momentary solution vector which just balances the current state of the evolving economic system. According to Kornai's terminology our understanding pertains to the "bookkeeping" notion of equilibrium.

The following Figure 19.5 illustrates our conclusions for the basic exchange framework with finitely many agents. Note that this time we use the symbol t = elapsing time for the evolution parameter.

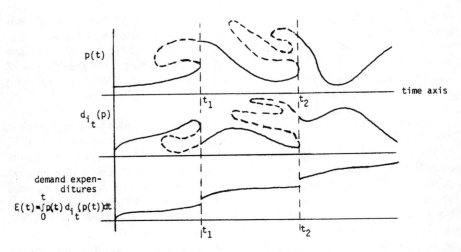

Fig. 19.5: ε-near Equilibrium Path with Backtracking Parts for a Market Evolution with Flow Commodities

Now what are the advantages of this result? The *first aspect* of yielding a homeostasis has already been discussed before. Secondly, the opportunity of adjusting equilibrium values along a (piecewise) continuous path during the evolution of the economic system means that searching for new equilibria of new states of the evolving economy is made maximally easy. In order to assess the *third* advantage let us look at it from the negative.[1] What do discrete, abrupt jumps of equilibrium values during an evolution over continuous time mean for the economic agents? Well, even though there appear different types of equilibrium values in the different basic models a general answer to this question is possible. A discrete abrupt jump of equilibrium values generally causes discrete abrupt changes of the economic status and behavior of the agents. Such abrupt changes are commonly disliked by economic agents for several reasons. First and foremost, they change the agents' economic status. Though this sometimes may be well beneficial to an agent, in reality one generally observes an attitude of risk aversion and conservatism in economic affairs. To be more specific: a discrete jump of equilibrium variables means in the basic models from the exchange framework of Chapters 4 and 5 that in general both the value of the initial endowments and the consumption in equilibrium change discontinuously. In the basic models from the framework with production, taxes, and subsidies of Chapter 6 in addition the expenditures by the government and the produced quantities change discontinuously. Finally, in the quantity constrained basic multi-sectoral model from Section 7.2 the equilibrium quantities of demanded and supplied types of labor and commodities change abruptly and discontinuously.

The widespread attitude of economic agents of risk aversion and conservatism is also the source of a second type of reservation about sudden discontinuous changes of equilibria. A sudden discontinuous change in independent equilibrium variables and in physical economic magnitudes may well create the expectation of further sudden discontinuous changes. Since the building of expectations and their intertemporal effects are per definitionem not explicitly modelled by our kinetic approach, we must here confine ourselves to the remark that expectations of unknown discontinuities of the economic process in general give rise to negative associations. This has to do with the speculative distortion of individual plans, and generally with the fear of destruction of the own economic status and of that of the community by destabilizing the economic processes. Empirical observations strongly support this. One may only think for instance of the international foreign exchange markets and stock markets, or the oil market. Following K. Galbraith it is even one of the crucial aims of the large international corporations, which form the 'planning system' in his terminology, to ensure a foreseeable smooth performance of the whole economic process.

---

[1] Thus we give a foundation of Balasko's statement (1988, p. 70): "... the idea that discontinuity is in itself harmful, synonymous of catastrophies (sic), is widespread ... We shall content ourselves with the idea that, from an economic point of view, a continuous evolution path is superior to any discontinuous one."

We finally want to emphasize that our result only ensures the *opportunity* to piecewise continuously adjust equilibrium values during any arbitrary continuous evolution of the economy. Remember that there is no uniqueness of a (near-) equilibrium path. Furthermore, to repeat it, we neither provide a theory of *which* equilibria from the evolving equilibrium set are actually selected, or *how* they are chosen, nor do we even claim that equilibrium states in a model actually would reflect the observed states of real economies. Our findings just give reference solutions in reference models.

19.2.2 The Frequency Model of a Piecewise Continuously Balanced Evolving Economy

In this Section we are going to provide a new approach of modelling and analyzing an evolving economy in continuous time. In our eyes it has the strong advantage of being close to reality. As in the preceding Section we consider ongoing economies which are *left to itself,* i.e. there is no agency which systematically would take any intervening measures. The basic intuition for what we will subsequently call our *frequency model of an evolving economy in continuous time* comes from asset chart analysis. As the reader may remember a tool from descriptive statistics which is frequently used in chart analysis are *curves of sliding averages.* (For instance, the curve of sliding 200-days averages of a certain share plots the arithmetic mean of the last 200 quotations over the horizontal time axis.) This will give us the basis for the following intuition.

In contrast to the discrete approach in Chapter 19.1, we do not string 'brick' periods along the time axis, but let an interval (a period) of fixed positive length β monotonically slide along the time axis. Considering the left boundary point t of the *sliding period* $[t, t + \beta]$ as the current date we may think of $[t, t + \beta]$ as a monotonically *sliding (forward) planning period,* i.e. as monotonically sliding time horizon of future planning. Now, in this setting we have the following natural notion of demand and supply over continuous time. Let us divide the length β of the sliding planning period at any date t by the aggregate planned number of demanded (supplied) units of quantity during $[t, t + \beta]$. In other words, we replace the usual measurement of demand and supply over discrete time by *units of quantity per unit time period* by its *inverse,* i.e. by *units of time passing between two successively demanded (supplied) units of quantitiy.* Thus a large number of planned demanded (supplied) units of quantity during $[t, t + \beta]$ at date t corresponds to a small number of units of time between two successively demanded (supplied) units of quantity.

Strictly speaking, in this way we design a *'wavelength model'* of an evolving economy over continuous time rather than a 'frequency' model, since in physics frequency characterizes the number of events per unit of time, whereas wavelength measures the distance. Nevertheless, since our basic intuition comes from the number of demanded (supplied) units of commodities during the sliding planning period we will adhere to the term 'frequency model'.

This intuition provides the basis for the following general formalization of an evolving economy in continuous time in the context of the exchange framework from Chapter 5. At any date t we denote by $\hat{d}_{i_t}(p)$ the "length of the time interval between two successive planned demanded units of quantity on the i-th market resulting from the individual plans of all agents valid for price vector p". This holds from time t on until further notice. Analogously, we define $\hat{s}_{i_t}(p)$ on the supply side of the i-th market.

This means, at any date t we have a t-state economy $((\hat{d}_{i_t}), (\hat{s}_{i_t}))_{i=1,\ldots,n}$. Let us now go one step further and let us consider the n difference functions $\hat{\zeta}_{i_t}(p) := \hat{s}_{i_t}(p) - \hat{d}_{i_t}(p)$. Recalling the economic meaning of $\hat{s}_{i_t}(p)$ and $\hat{d}_{i_t}(p)$ it becomes immediately clear that $\hat{\zeta}_{i_t}(p)$ actually expresses excess demand even though this time the demand term $\hat{d}_{i_t}(p)$ is subtracted from the supply term $\hat{s}_{i_t}(p)$. In fact, $\hat{\zeta}_{i_t}(p)$ is greater than 0 when the planned time interval between two successively supplied units of commodity i is greater than the planned time interval between two successively demanded units of commodity i. But economically this just means that demand is in excess of supply! Clearly, p_t^0 is a momentary equilibrium for the t-state economy $((\hat{d}_{i_t}), (\hat{s}_{i_t}))_{i=1,\ldots,n}$ if $\hat{d}_{i_t}(p_t^0) = \hat{s}_{i_t}(p_t^0)$ for $i = 1, \ldots, n$, i.e. if $\hat{\zeta}_{i_t}(p_t^0) = 0$ for all i. Thus, we can also say that the equilibrium price vector p_t^0 "phases" planned demand and supply on all markets at date t.

In addition, let us assume that the functions $\hat{d}_{i_t}(-)$ and $\hat{s}_{i_t}(-)$ have such properties that the t-state excess demand system, or say the t-state economy, $(\hat{\zeta}_{i_t})_{i=1,\ldots,n}$ has the properties of our basic exchange model without Walras' law and homogeneity of Section 5.1. (We may equally well require that it fulfills the generalized properties from Section 5.2.) This is clearly a reasonable assumption since Walras' law evidently is completely meaningless for our frequeny approach. We furthermore call it to the reader's mind that the assumptions of our basic models from Chapter 5 are considerably weak.

As in the preceding Section we again are in the fortunate position to have all analytical results from Part II at our disposal. Thus we can state that we have provided a frequency model of a piecewise continuously balanced evolving economy in continuous time. More detailed, the existence of (near-) equilibrium paths means the following: in our frequency set-up for any evolving economy in continuous time it is possible to permanently phase, or say synchronize, demand and supply events on each market by piecewise continuously adjusting equilibrium prices. For the economic achievements of this conclusion the reader is referred to the discussion at the end of Section 19.2.1.

In order to further support the reader's intuition let us give the following illustration (see Figure 19.6). For each of the n markets of the economy imagine a screen where on two horizontal axes, one below the other, the successive demand and supply events over time are recorded as light points moving with constant velocity from the right boundary of the screen to the left. Moreover, any 'market screen' is bisected by a vertical line in the middle whose intersection points with the two horizontal axes mark the *present* moment. To the *right* of the vertical middle line on the two horizontal axes *planned* future demand and supply intervals under the prevailing momentary price vector are visualized. On the left sides we see the intervals which actually have been realized up to the present moment. When prices are adjusted over time following an equilibrium price path, then both the realized demand and supply events on the left sides are phased and particularly also the planned events on the right sides. Clearly, the length of the planned intervals on the right sides of the two horizonal axes is of constant length, i.e. planned demand and supply events are equidistant 'until further notice'. The realized intervals, however, are in general not of constant length when the economic behaviour during the evolution changes and equilibrium prices are adjusted.

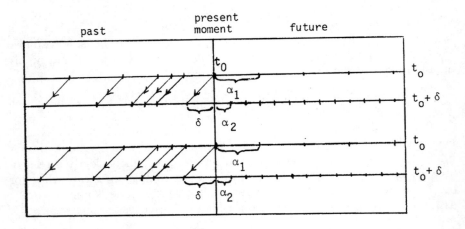

Fig. 19.6: Realized and Planned Demand and Supply in an Evolving Market

Figure 19.6 shows realized and planned demand and supply on market i at the two dates t_0 and $t_0 + \delta$. At t_0 and equilibrium price vector p_{t_0} the planned length of demand and supply intervals is α_1, whereas at $t_0 + \delta$ demand and supply plans have changed such that under the new equilibrium price vector $p_{t_0+\delta}$ the planned common length of demand and supply intervals is now $\alpha_2 < \alpha_1$.

To make the reciprocal relationship between the conventional way of formalizing demand and supply and our present approach more intuitive we provide the following illustration. First let us introduce the following technical monotone one-to-one function (see Figure 19.7)

$$\varphi : [0, \infty[\longrightarrow]0, \infty[$$

with the properties

$$\varphi(0) = 1 + 1/\epsilon$$
$$\varphi(\epsilon) = 1/\epsilon$$
$$\varphi(x) = 1/x \quad \text{for} \quad x \geq \epsilon$$

where ϵ is an arbitrarily small positive real number.

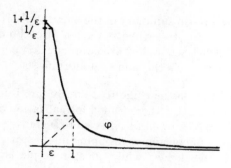

Fig. 19.7: Monotonic One-to-One Auxiliary Function

φ will serve as a transformation rule for translating magnitudes of quantity into time intervals. Thus let us define

$$\hat{d}_{i_t}(p) := \varphi(d_{i_t}(p))$$
$$\hat{s}_{i_t}(p) := \varphi(s_{i_t}(p))$$

where $d_{i_t}(-)$ and $s_{i_t}(-)$ are the conventional t-state demand and supply function for the i-th commodity respectively. (Note that we did not explicitly introduce the functions $d_{i_t}(-)$ and $s_{i_t}(-)$ in the general exposition of our basic exchange models without Walras' law and homogeneity in Chapter 5. Nevertheless, there is no difficulty to conceive of functions $d_{i_t}(-)$ and $s_{i_t}(-)$ with reasonable properties such that the derived conventional excess demand function $\zeta_{i_t}(p) := d_{i_t}(p) - s_{i_t}(p)$ fits into the mentioned models.)

Clearly, our definition is reasonable since φ transforms a *large* number of units of quantities which are planned for demand/supply during the period

$[t, t+\beta]$ into a *small* time interval between any two successive demand/supply events, and vice versa. The reader will have noticed that, in order to yield a continuous monotone transformation φ which is defined on the whole $\mathbb{R}_+$, we have associated an arbitrarily large "artificial" interval length $\frac{1}{\epsilon}+1$ with zero demand/supply. Obviously, p^0 is an equilibrium price vector for the t-state economy $((\hat{d}_{i_t}), (\hat{s}_{i_t}))_{i=1,\ldots,n}$ if *any pair* $(\hat{d}_{i_t}(p^0), \hat{s}_{i_t}(p^0))$, $i = 1, \ldots, n$, is "*phased*". Equivalently, $\hat{\zeta}_{i_t}(p^0) := \hat{s}_{i_t}(p^0) - \hat{d}_{i_t}(p^0) = 0$ for $i = 1, \ldots, n$.

So we can summarize: since φ is one-to-one with strictly positive values, p^0 is an equilibrium price vector of the t-state economy $((\hat{d}_{i_t}), (\hat{s}_{i_t}))_{i=1,\ldots,n}$ *if and only if* it is a zero of the excess demand system $(\zeta_{i_t})_{i=1,\ldots,n}$ in the usual sense with $\zeta_{i_t}(p) = d_{i_t}(p) - s_{i_t}(p)$. Or, in other words: The formal equilibrium analysis for the introduced frequency model is all the same as for the exchange model without Walras' law and homogeneity from Section 5.2.

Figure 19.8 below gives a summary of the *qualitative properties* of the functions $\hat{d}_{i_t}(-), \hat{s}_{i_t}(-)$, and $\tilde{\zeta}_{i_t}(-)$ when the generalized boundary assumptions of the model version from Section 5.2 are adopted. Comparing ζ_{i_t} with $\hat{\zeta}_{i_t}$ one sees that both functions have the same equilibrium set.

After this illustration let us come to a concluding brief discussion of our frequency approach. Actually, the outlined formalization of demand and supply over continuous time can well be assessed to be fairly realistic. The reader may think for instance of consumers dropping into a store and taking cans, or something else, from the shelves, while the storekeeper is putting the articles into the shelves. Or one may think of successive orders for a certain type of car written into the sellers' order book, and of the line of produced cars leaving the factory. States of disequilibrium, i.e. of unphased sequences of demand and supply events, manifest by queues of byers or by compulsory stocks.

Admittedly, our formalization means a certain stylization of the real performance of demand and supply over time. In fact, agents do not always demand or supply only one unit of a commodity at one date. However, the intuition behind the sliding period lets appear our formalization as a relatively harmless stylization of reality.

19.3 Time Consuming Equilibrium Adjustment Processes

Looking more closely at the flow model from Section 19.2.1 one is naturally led to a problem which has already been addressed in the early days of mathematical economics. In a paper from 1930 Rosenstein-Rodan emphasized that someone who is adjusting non-equilibrium values to equilibrium ones in a

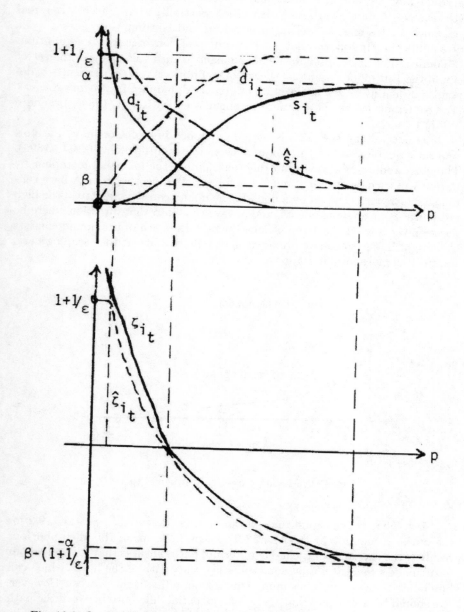

Fig. 19.8: Qualitative Properties of the Functions of the Frequency Model
$\hat{d}_{i_t} = \varphi(d_{i_t})$, $\hat{s}_{i_t} = \varphi(s_{i_t})$, $\zeta_{i_t} = d_{i_t} - s_{i_t}$, $\hat{\zeta}_{i_t} = \hat{s}_{i_t} - \hat{d}_{i_t}$, p^0 is the p-axis-coordinate of the tangential point of ζ_{i_t} and $\hat{\zeta}_{i_t}$ and at the same time the p-axis-coordinate of the intersection point of d_{i_t} and s_{i_t}.

gradual way, i.e. not in infinitesimal time, in general faces a moving target (Rosenstein-Rodan, 1930). As Fisher (1983, Section 1.6) has put it: "In a real economy ... trading, as well as production and consumption, goes on out of equilibrium. It follows that, in the course of convergence to equilibrium (assuming that occurs), endowments change. In turn this changes the set of equilibria. Put more succinctly, the set of equilibria is path dependent – it depends not merely on the initial state but on the dynamic adjustment process". (For contributions which are more technical see also Kloek (1984), Legendre (1987).)

It is noteworthy that this issue has a famous predecessor in V. Pareto's "courbes de poursuite" (in Cours d'economie politique, § 40, 41, 1896). Though standing in a somewhat different context the following metaphor by Pareto will turn out to be of much use for expounding our issue. A hare running along a wall is chased by a hound. The hound can see the hare in the high grass merely at a few moments, and always runs into that direction where he has seen the hare the last time. Consequently, the path of the hound consists of straight line segments approximating the linear path of the escaping hare. Figure 19.9 gives an illustration.

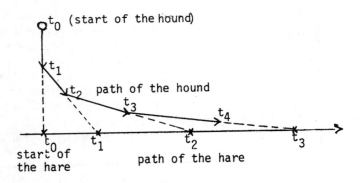

Fig. 19.9: Moving Target Adjustment Process

Now, what is the significance of our results presented in Section 19.2 for the outlined 'hound-hare-problem'? To speak in terms of this metaphor our result of the existence of (near-) equilibrium paths ensures that for each of our basic models *there is always* something like the "path of the hare" which can be "hunted" by a time-consuming equilibrium adjustment process. However, one should be well aware that there is an essential difference between the path of a hare and a (near-) equilibrium path. The path of a hare is monotonically one-parametrized by elapsing time, whereas a (near-) equilibrium path may backtrack with respect to the time parameter.

This leads us to the following *summary*: for a time-consuming equilibrium adjustment process a (near-) equilibrium path provides the opportunity to

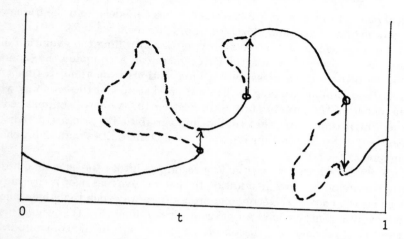

Fig. 19.10: Moving Target Adjustment with Jumps at Backtracking Parts

aim at a *target* which moves continuously - up to finitely many discrete jumps at critical states of the underlying evolution of economies. Figure 19.10 gives an illustration with the real time parameter t moving along the horizontal axis.

Following the backtracking parts of the (near-) equilibrium path, which are dotted in the example of Figure 19.10, leads to the targets of the unavoidable jumps. In addition, in Section 10.2 below the reader finds an algebraic criterion for identifying suitable targets for the jumps.

19.4 Frictionless Tuning of Coordination Signals in Evolving Economies in Continuous Historical Time

19.4.1 General Conceptualization

In Section 19.2 we have formalized and analyzed evolving economies over continuous time which evolve "openly", i.e. which are "left to themselves". Speaking in the terms introduced in Chapter 12, we have exclusively dealt with *course evolutions*. Nevertheless, in reality one frequently observes regulatory interventions by political-economic institutions.

Looking more closely at our basic models from the viewpoint of possible external control we find that three of our basic models which suit for the modelling over continuous time contain parameters can, at least in principle, be controlled by the government. These are the quantity constrained multi-sectoral model from Section 7.2 and the two basic models from the framework with production, taxes, and subsidies from Chapter 6.[2]

Generally speaking there are two types of controlling the evolution of an economic system possible: first the government can accompany the economic process with discretionary regulating interventions from time to time. And second, it can aim at reaching a certain desired state b of the economy which is different from the present state a. In Section 19.2 we have summarized the reasons why it is desirable that equilibrium variables, i.e. prices, tax redisbursals, subsidy rates, and rationed quantities of commodities and labor, change continuously.

In Section 19.2 we have ensured by assumption that the economic process evolves continuously. Now, in addition to that we give reasons why it is advantageous also to control the economic process in a way such that the economic system evolves continuously. As before let us assume that this would not be the case, i.e. that there are discontinuous finite leaps in the control parameters and hence in the evolution of economic states. Then in general one will face two categories of costs, namely (1) administrative costs, and (2) political costs. As Hettich (1979) puts it in the context of tax reform: "Obstacles to the adoption of the preferred tax base have proven too great in the past. Political constraints and high administrative cost impede reformers. As a result, partial improvement are a more interesting and more relevant concern for analysis" (see p. 695). And Hettich continues (p. 706): "Alterations in the tax system are not free, however. Real resources must be spent in order to accomplish such change. The cost of reform can be grouped into *two broad categories*. The *first* one may be labelled administrative costs. It includes all the expenditures of real resources associated with introduction of new tax forms, changes in reporting, collection, and enforcement. While these costs are generally small in relation to the revenues collected, they may become large for certain income components included in the Haig-Simons concept of income. The *second* type of costs is more difficult to define. The term political costs suggests some of the connotations, although it is misleading in other respects. Perhaps it is best to refer to an actual case. When the government of Canada in response to the Report of the Royal Commission on Taxation started the process of tax reform, it committed to that end much time and effort by both bureaucrats and political leaders. These resources could have been spent to achieve (or at least debate) different social aims or to enact other programs increasing justice or equity in Canadian society."

This view was also at the basis of the 'fine tuning' policy of the Nixon administration and of the following famous statement by Woodrow Wilson which is quoted by Feldstein (1976, p. 77): "We shall deal with our economic system as it is and as it may be modified, not as it might be if we had a

---

[2] We are dealing with these three models in this order since the quantity constrained multi-sectoral model from Section 7.2 provides the economic and formal intuition for the two other models from Chapter 6.

clean sheet of paper to write upon; and step by step we shall make it what it should be." Feldstein resumes (loc. cit.): "Optimal tax reform must take as its starting point the existing tax system and the fact that actual changes are slow and piecemeal."[3]

In the eighties there has grown another branch of literature which also addresses the topics of continuous modellation of evolving economic systems and of gradual versus "bang-bang" control by an agency (see e.g. Gandolfo/Padoan (1984, esp. Sections 1.2 and 4.5), Gandolfo/Petit (1988)). As to the economic advantages of a continuous approach to model an evolving economic system the authors primarily emphasize the "smoothing"[4] effect over time of aggregating many individual operations (cf our second epigraph to this Chapter), and hint to the awkward indeterminateness problem with the length of the 'period' (see Gandolfo/Padoan (1984, Section 1.2)). The question whether a policy strategy should be implemented by a gradual, i.e. continuous, tuning of control variables, or rather by "bang-bang" or "cold turkey", i.e. a shock therapy, is treated by means of comparative simulation scenarios by Gandolfo/Padoan (1984, Section 4.5). In Fellner et alii (1981) one finds a general discussion on the controversy shock therapy vs. gradualism.

The main result of this Section will be that in our three basic set-ups addressed above a *frictionless control over continuous time* is always possible *for any course or conncetion evolution* – if the economic behavior which is not governed by the control parameters remains unchanged.

Definition 19.1. *The term "frictionless change (or tuning, control)" means that all control parameters, economic states, and (near–) equilibrium vectors change continuously.*

To fix ideas let us start with a unifying re-formalization of a continuous evolution of economies in each of the three addressed basic models[5]. Let us formally represent an evolution by a composite continuous mapping

$$[0,1] \xrightarrow{z} \quad C_h \quad \xrightarrow{\Phi_z^h} \mathcal{E}_h$$
$$s \mapsto (c_{1_s}, \ldots, c_{r_s}) \mapsto E_{(c_{1_s}, \ldots, c_{r_s})}$$

where $\mathcal{E}_h$, $h = 1, 2, 3$, denotes the space of economies of the basic model h, and C_h the Euclidean subspace of control parameters of model h. The symbol z means any continuous path in the control parameter space C_h, and Φ_z^h is a fixed mapping which, nevertheless, may be individually chosen in dependence

---

[3] There is a broad literature on piecemeal tax reform. See for instance also Guesnerie (1977), or Hatta (1977). Zodrow (1981) sums up some ciriticisms to this approach. For a comprehensive survey see Atkinson/Stiglitz (1980), Chapter 12.

[4] We use here the term "smooth" in the colloquial sense and do not refer to its meaning in the differentiability sense.

[5] The Subsections 19.4.2 and 19.4.3 below will provide the specifications of the general formalization to the addressed three basic models.

on the evolution of the control parameter path z. More specifically, Φ_z^h associates an economy $E_{(c_{1_s},\ldots,c_{r_s})}$ with any admissible control parameter tuple in a *continuous* way. Consequently, any admissible evolution is completely characterized by the evolution (the path) z of control parameters. Again, all analytical results from Part II are valid.

To illustrate it we can write the vector of control parameters $(c_{1_s},\ldots,c_{r_s})$ on any s-slice of the homotopy space $H_h \times [0,1]$ as it is illustrated in Figure 19.11 below. For the example of Figure 19.11 a frictionless control obviously

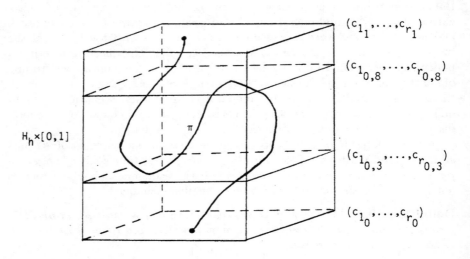

Fig. 19.11: Frictionless Control in the Homotopy Space

can be achieved by following the (near-) equilibrium path π during the part from $(c_{1_0},\ldots,c_{r_0})$ to $(c_{1_{0,8}},\ldots,c_{r_{0,8}})$, then running back in the control parameter path from $(c_{1_{0,8}},\ldots,c_{r_{0,8}})$ to $(c_{1_{0,3}},\ldots,c_{r_{0,3}})$, and finally running forward again from $(c_{1_{0,3}},\ldots,c_{r_{0,3}})$ to $(c_{1_1},\ldots,c_{r_1})$. Extending this observation to the general case we can sum up: a frictionless control of an evolution $\Phi_z^h \circ z$ in any of the addressed three basic models is possible by an appropriate continuous *re-parametrization* $\widetilde{z}$ of the control parameter path z. Formally $\widetilde{z}$ is obtained by projecting the (near-) equilibrium path π on the space $[0,1]$ of the homotopy space $H_h \times [0,1]$, i.e. $\widetilde{z} = pr_2 \cdot \pi$. Hence, both paths, the originally chosen path $\pi : [0,1] \to H_h \times [0,1]$ of (near-) equilibria and the appropriately re-parametrized path $\widetilde{z} : [0,1] \to z([0,1]) \subset C_h \subset \mathbb{R}^l$ are geometrically nicely behaved.

At this point arises a natural question: Where can we find the historical time t in this set-up? Do we have to become "younger" in order to accomplish frictionless tuning in this context? Obviously, the answer to the last question is "no" since one has to identify the historical time t with the evolution parameter of the re-parametrized evolution $\Phi_z^h \circ \tilde{z}$. Thus, the vector of control parameters at time t is $\tilde{z}(t) = pr_2(z(t))$, and the state of the economy at time t is $\Phi_z^h(\tilde{z}(t)) = \Phi_z^h(pr_2(z(t)))$. This is completely analogous to the situation of a film which is played with several parts backtracked and then played forward again. Viewers (economic agents) are not getting younger when the film (the evolution) is backtracking, but the whole showing is expanded by the length of time the repeated parts require.

Our findings so far lead to the following general conclusions. We consider evolutions of economic states in any of the addressed three basic settings which are either "left to themselves", i.e. which are openly evolving *course evolutions,* or are a priori chosen by some exogenous political-economic agency such that a certain desired state of the economy is approached *(connection evolutions).* In any way an evolution of economic states is completely determined by the associate evolution of control parameters. Now, if the political-economic agency aims at a completely frictionless control in the described sense it generally has to move back and forth in the evolution of control parameters. This is due to the backtracking feature of (near–) equilibrium paths. This means that in the present basic set-ups in general it is inevitable for an (exogenous) political-economic agency – if it is purposed to ensure a frictionless control – to give the impression of somewhat being undecided in choosing appropriate measures.

In the following two subsections we will provide the due specifications of this general exposition to each of the three basic set-ups addressed above. (Please note Footnote 2 above concerning the order of model frameworks in Sections 19.4.2 and 19.4.3.)

19.4.2 Frictionless Tuning in the Quantity Constrained Multi-Sectoral Model From Chapter 4

The specification of our general conceptualization above to the quantity constrained multi-sectoral model from Section 7.2 is straightforward. There is only one vector of control parameters, namely the $2m$-vector of prices and wages in the m sectors. Thus, the space of control parameters is $C_1 = \mathbb{R}_+^{2m} \setminus \{0\}$.

Let us assume that there is an authority capable of extraneously controlling prices and wages. Then any continuous evolution in the present basic set-up – be it of the course or of the connection type – can uninterruptedly be tuned in a frictionless way following a geometrically nicely behaved path in the space of equilibrium variables C^{2m}. This is achieved by accordingly continuously reparametrizing the control parameters within the path describing the evolution. We again recall it that (near–) equilibrium paths in general are neither locally unique, nor do we explicitly model a selection mechanism. Our position is that of ensuring the opportunity of exerting a frictionless control as described.

19.4.3 Frictionless Tuning in the General Equilibrium Framework With Production, Taxes, and Subsidies From Chapter 3

In order to specify the general exposition given in 19.4.1 to the first basic model with production and taxes from Section 6.1 let us confine us to the example of consumption and income tax schemes presented in Section 6.1. Recall that there are n commodities, $m > n$ production processes, and h economic agents in this example. There are *four vectors* of control parameters: the $n \cdot h$ - vector of individual ad-valorem consumption tax rates $(\sigma_{ij})_{\substack{i=1,\dots,n \\ j=1,\dots,h}} \in \mathbb{R}_{+}^{n \cdot h}$, the h-vector of individual endowment income tax rates $(\rho_j)_{j=1,\dots,h} \in [0,1[^h$, the h-vector of individual share rates of tax revenue $(\vartheta_j)_{i=1,\dots,h} \in \overline{\Delta}^{h-1}$, and the $n \cdot m$ - vector of ad-valorem production tax rates $(\tau_{ij})_{\substack{i=1,\dots,n \\ j=1,\dots,m}} \in [0,1]^{n \cdot m}$.

Clearly, all of these parameters are in principle amenable to control by some governmental authority. Consequently, we are facing the Euclidean control parameter space

$$C_2 := \mathbb{R}_{+}^{nh} \times [0,1[^h \times \overline{\Delta}^{h-1} \times [0,1]^{n \cdot m} \subset \mathbb{R}^{(nh+h+h+nm)} = \mathbb{R}^{(nh+2h+nm)}.$$

The extension of this framework with consumption and income tax schemes by *subsidy schemes* in the *second* basic model from Section 6.2 adds *three* further control parameter vectors: the $n \cdot h$-vector of the individual ad-valorem subsidy rates on final demand $(\delta_{ij})_{\substack{i=1,\dots,n \\ j=1,\dots,h}} \in [0,1]^{nh}$, the h-vector of individual income transfers to the11.4 consumers $(\eta_j)_{j=1,\dots,h} \in \mathbb{R}_{+}^{h}$, and the $n \cdot m$-vector of ad-valorem subsidy rates on production $(\chi_{ij})_{\substack{i=1,\dots,n \\ j=1,\dots,m}} \in [0,1]^{nm}$.

11.0Hence, the space of control parameters is now

$$C_3 := C_2 \times [0,1]^{nh} \times \mathbb{R}_{+}^{h} \times [0,1]^{nm} \subset \mathbb{R}^{(nh+h+h+nm+nh+h+nm)} = \mathbb{R}^{(2nh+3h+2nm)}.$$

As to the question of the possibility of a frictionless tuning the same conclusions apply as in Subsection 19.4.1 with the only difference that now the space of equilibrium variables is $\Delta^{n-1} \times \mathbb{R}_{+}$ (or $\Delta^{n-1} \times \mathbb{R}_{+} \times [0,1]$).

20
Conclusions

In Part III, we have applied the analytical results from Part II to the frameworks of economic evolutions in Part I. We have seen that in the field of atemporal economic problems the results allow for significant extensions of two established methods which, in their traditional form, have come to a dead end. These two methods are the path following method for the computation of equilibria and the comparative statics method. For the path following method our results achieve an extension for computing equilibria of non-regular economies. Comparative statics is given a new meaning in the case of multiple equilibria by our results.

The main applications of Part III, however, pertain to evolutions of economies in historical time. We have first clarified the scope of our investigation. It is not that of traditional dynamic analysis which is intended to explain the causal intertemporal relationships of successive states of an evolving economy. Rather we search for general regularity properties of the dependent evolution of the endogenous equilibrium values that hold true for any admissible evolution of the economy. Thus, our approach can well be seen as complementary to the recent branch of evolutionary economics which strives for a new 'open' approach to the explanation of evolutions of economic systems.

Chapter 19.1 analytically conceptualizes ongoing economies in discrete historical time and presents some applications of our extensions of the path following method and of comparative statics. The main body of our applications, however, is based on modelling economies evolving in continuous historical time. Starting from our concepts of economic evolutions given in Part I, we first provide two alternative new ways to analytically conceptualize an evolving economy in historical continuous time. Our main results ensure the opportunity for an exogenous agency to exercise a continuous 'double fine tuning' of both evolutions, that of equilibrium variables and of economic state parameters. From a formal viewpoint the opportunity of adjusting equilibrium values along a continuous path during an evolution of the economy has the undisputable advantage of making it easy to find new equilibria for new states of the economic system. The pros and cons of a continuous tuning policy versus discontinuous cold turkey policy interventions have been discussed extensively in the 1970ies debate on piecemeal versus shock therapy tax reform, and later in the macroeconomic optimal policy design controversy.

In a nutshell, our findings are the following. For some of the basic models in Part I which contain explicit control parameters, a truly permanently continuous, i.e. a frictionless, gradual double tuning of equilibrium values and of economic state parameters is possible within any given evolution of the economic system. However, in general, it is necessary for the controlling agency to backtrack partly along the path of state parameters, otherwise, discrete jumps in the equilibrium values are generally unavoidable, as simple examples show. In other words, the controlling agency can only achieve a frictionless double tuning at the price of giving the impression of being somewhat undecided in its controlling activities.

General Conclusions and Outlook

General Conclusions and Outlook

General Conclusions and Outlook

The cornerstones of the present book are given by two key concepts of economics: equilibrium and evolution. At first glance, these concepts seem to be contrary, even irreconcilable: while equilibrium usually stands for a final state of rest of the economic system, evolution means movement and development by change. Following this common understanding it is only consequent to see modelling approaches employing the equilibrium concept in strict rivalry to modelling approaches devoted to the idea of evolution. However, a trickling question remains: How can two theoretical key concepts that explore the same subject, i.e. the economy be essentially contrary?

A first answer to this question may be the general remark that rivalling theoretical positions in the history of the social sciences have always helped to trigger creative new ideas. But this answer appears not to be satisfactory. In fact, both concepts of evolution and equilibrium enhance our understanding of economic systems. Accordingly, they should not appear as mutually exclusive concepts to analyse economies, but rather as complementary ones. One could even see them as coexistent features of real economic systems. Thus, endeavours to reconcile, or synthesize, the two concepts in a manner which accounts for the essentials and the peculiarities of both concepts promise to be of greater usefulness for economists than the usual statements about their disparateness and exclusiveness.

Accordingly, the purpose of the present book is to contribute to this program of bridging the gap between the two concepts of equilibrium and evolution. It starts from the idea that an equilibrium in an appropriate understanding should not be a state of final rest as it is in the case of thermodynamics. Surely the thermodynamic understanding of equilibrium would contradict any idea of evolution. Rather, an equilibrium in our context denotes a "bookkeeping equilibrium" in the sense of a momentarily balanced state of the evolutionary economic system under consideration. The best intuition for that is provided by the metaphor of a high wire performer who during his walk is in equilibrium at every single moment by continuously adapting the balance bar to his changing positions on the high wire.

To achieve its aims the present study provides a formal conceptualization of two crucial elements: first the independent evolution of an economic system, and second the dependently shaped evolution of momentary equilibria. We have carried out the first task in Part I of this book, and the second one in Part II. Speaking in terms of the preceding metaphor we are then looking at Part III for the question of how the performer has to adapt his balance bar. Or to put it into a question: Can he always achieve his aim of staying balanced while walking on the wire by continuously adapting the bar without any sudden movements, or does he have to perform wild and irregular movements and leaps with the bar? Speaking in terms of our metaphor, our findings are: In general the latter cannot been avoided completely if the performer only progresses forward on his wire, but it can be reduced to certain "critical" points on the performer's path. If the performer is prepared to backtrack from such critical points on his wire, and afterwards moves forward again on the wire with a suitably different adaptation of his bar he can avoid all sudden leaps in the bar's movement.

If the reader wants to re-translate this metaphor into the context of the present analysis she can do so by considering the following points:

- the high wire corresponds to the evolution of the economic system modelled by any of the frameworks in Part I of this book,
- the performer's momentary position on the high wire corresponds to the economy's momentary position during its evolution, and
- the balance bar corresponds to the coordination signal vectors, i.e. to the equilibria.

But what about the re-translation of the usefulness of a continuous adaptation of the balance bar into the economic context? Surely, for the high wire performer it is a matter of survival whether he adapts his balance bar properly, or not. But mutatis mutandis it is also in two ways useful for the economist to know about the existence of a "(near) equilibrium path". First, in a purely formal sense it helps to compute equilibria and to apply comparative statics in the case of multiple equilibria. Second, a (near) equilibrium path makes possible what throughout our study is called a frictionless, i.e. non-disrupted, continuous tuning of coordination signals, i.e. equilibria, in open loop evolving economies.

Let us conclude with an outlook on the relevance of the findings for evolving market systems in reality. To simplify things let us think of a real stock market. The first question is why we do not observe piecewise, or completely continuous equilibrium price paths during the evolution of a stock market. A first answer to this question could be the rejection of the question itself because it contains an incorrect hypothesis. Indeed, as one might argue, we do observe piecewise continuous equilibrium paths on stock markets: crashes indicate the sudden leaps of equilibria. However, let us leave this somewhat tricky controversy and let us look at the real market system evolution of a stock market more closely. How are prices formed on a real stock exchange?

Observations of real price forming processes on stock markets strongly support the hypothesis that prices of stocks are not formed simultaneously and interdependently by real agents. Rather, the agents find them in a "parametrical" manner, i.e. sequentially under a ceteris paribus condition. By that we mean that the equilibrium price of each individual stock is formed as a momentary market clearing price given the perceived actual transaction prices, i.e. momentary equilibrium prices, of all other stocks at that moment. In this way the agents on a real stock market circumvent the problems coming from the simultaneous and fully interdependent formation of all equilibrium prices.

However, this argument only applies to the case of "continuous trade", not to single, or once-a-day, auctions (producing a "cassa price" for each stock). In a once-a-day auction stock prices are formed over a period of time, not at a moment, and thus involve some sort of "tâtonnement" process of the price formation by the stock market agents. These real "tâtonnement" processes, however, might also not be ruled by the strict theoretical assumptions for tâtonnement processes. Instead they might be characterized by the fact that real agents on the stock exchange floor can re-shape the excess demand functions in a way that simplifies the finding of equilibrium solutions. This is not to be misunderstood as a demand or supply manipulation, but it is a consequence of the brokers' professional obligation "to smooth" stock price evolutions "according to the real market situation".

At this point the reader might ask a question that is most familiar to a theoretical economist, namely, what theoretical concepts and analyses are useful if real agents and markets do not behave like the theoretical concepts and results claim. Of course, an exhaustive answer to this fundamental question would be unacceptably lengthy. For instance, one would have to employ the whole arsenal of epistemological and methodological arguments dealing with the virtues and merits of the (mathematical) modelling approach in economics. As a first brief answer, however, we might argue that the more real economies develop their information and telecommunication devices, the more they will fulfil the requirements of theory. However, this argument does not completely answer the former question. Since we do not want here to overstretch the reader's patience, we confine ourselves to a practical argument forwarded in particular by Kenneth Arrow in his writings. In a nutshell Arrow's argument says that a theoretical framework should not solely be judged from the degree of its factual coincidence with the empirical evidence. Though it might not properly fit the real subject, it is meant to describe and to explain a theory or a model, in fact, it may achieve valuable insights into the nature of its subject. Considering the discrepancies between a theory and its real subject may well provide a sound basis for a theorist to develop more appropriate theoretical approaches in the future, which hopefully better fit their real subjects.

Appendix A[1]

Proof of Theorem 11.8

Possibly the following proof might appear somewhat lenghty to the reader. Nevertheless, it is based on just a few central ideas and follows similar lines as our constructions in Section 11.1.1. The proof divides up into two main steps.

(I) Let us begin with an evolution $(\zeta_s, t_s, A_s, A_s^*)_{s \in [0,1]}$ in the basic set-up with production and tax schemes from Section 6.1. In part (II) of the proof below we then will point out the modifications needed for evolutions from the basic set-up with production and tax and subsidy schemes from Section 6.2. In the sequel we will use the form $\zeta : \Delta^{n-1} \times \mathbb{R}_+ \longrightarrow \mathbb{R}^n$ and $\zeta : \Delta^{n-1} \times \mathbb{R}_+ \times [0,1] \longrightarrow \mathbb{R}^n$ respectively for the representation of excess demand functions.

Actually, we will show more than the Theorem alleges, namely that there is a $\beta_0 > 0$ such that for any $\epsilon > 0$ there is an evolution $(\hat{\zeta}_s, \hat{t}_s, \hat{A}_s, \hat{A}_s^*)_{s \in [0,1]}$ in our basic set-up which ϵ-uniformly approximates the given one such that the equilibrium sets of both evolutions are even contained in the compact set $W := (\Delta^{n-1} \times [0, \beta_0]) \times [0,1]$.

The proof of this statement will proceed *in three steps*. In the *first* step we will check that there is a $\beta_0 > 0$ such that the equilibrium set of the given evolution is contained in W. In the *second* step we will construct ϵ-uniformly approximating one-parametrizations $(\hat{\zeta}_s)_{s \in [0,1]}, (\hat{t}_s)_{s \in [0,1]}, (\hat{A}_s)_{s \in [0,1]}$, and $(\hat{A}_s^*)_{s \in [0,1]}$, and will show that the equilibrium set of the approximating evolution is also contained in W. And in the final *third* step we will demonstrate that the equilibrium set of the approximating evolution has 'nice', i.e. finitely piecewise analytical, equilibrium paths (see 'Mathematical Preliminaries').

---

[1] The author thanks Franz Bilitewski for substantial help and comments in the proof of Theorem 11.8.

However, before we can start with the main line of our proof we need some further concepts from geometrical and algebraic topology which will turn out to be most useful in the sequel.

First, we need a certain generalization of the notion of semi-algebraic subsets of $\mathbb{R}^n$ (see Section 11.1.1.). To this end we introduce the following

Definition 22.1. *Let $U \subset \mathbb{R}^n$ be an open subset and $\mathcal{F}$ a set of continuous real-valued functions on U. For any $A \subset U$ we say that $\mathbf{A}$ is described by $\mathcal{F}$ if there are functions $f_{ij}, g_{ij} \in \mathcal{F}$ such that*

$$A = \bigcup_{i=1}^{k} \bigcap_{j=1}^{l} (A_{ij} \setminus B_{ij}),$$

where $A_{ij} := \{x \in U | f_{ij}(x) > 0\}$ and
$B_{ij} := \{x \in U | g_{ij}(x) > 0\}$.

Recall that a function $f : U \longrightarrow \mathbb{R}$ from an open subset $U \subset \mathbb{R}^n$ is called *real analytical* iff for any $z_0 \in U$ there exists an $\epsilon > 0$ and a power series P which is absolute convergent on $\{z \in \mathbb{R}^n | \|z - z_0\| < \epsilon\}$ such that $f(z) = P(z - z_0)$ for any z from this set. In other words, a real analytical function is locally represented by power series.

Furthermore, let us introduce the

Definition 22.2. *For an open subset $U \subset \mathbb{R}^n$ and a $y \in \mathbb{R}^m$ we define*

$$\mathcal{M}(\mathbf{U}) := \{f : U \longrightarrow \mathbb{R} | f \text{ is real analytic}\}$$

and $\mathcal{M}(\mathbf{U})[y_1, \ldots, y_m] := \{g : U \times \mathbb{R}^m \longrightarrow \mathbb{R} | g(x, y)$
$= \displaystyle\sum_{(\nu_1, \ldots, \nu_m) \in \mathbb{N}^m} h_{\nu_1, \ldots, \nu_m}(x) \cdot y_1^{\nu_1} \cdot \ldots \cdot y_m^{\nu_m}$ such that all $h_{\nu_1, \ldots, \nu_m} \in \mathcal{M}(U)$
and $h_{\nu_1, \ldots, \nu_m} \equiv 0$ for all but finitely many m-tuples $(\nu_1, \ldots, \nu_m)\}$.

From the definition follows immediately that $x \subset \mathbb{R}^n$ is semi-algebraic iff x is described by $\mathbb{R}[x_1, \ldots, x_n]$. Now let us generalize this notion by the following

Definition 22.3.
(i) We call $X \subset \mathbb{R}^n$ semi-analytical iff for any $a \in \mathbb{R}^n$ there is an open neighborhood $U(a)$ of a in $\mathbb{R}^n$ such that $X \cap U(A)$ is described by $\mathcal{M}(U(a))$.

(ii) We call $X \subset \mathbb{R}^n \times \mathbb{R}^m$ $\mathbb{R}^m$-semi-algebraic-analytical iff for any $a \in \mathbb{R}^n$ there is an open neighborhood $U(a)$ of a in $\mathbb{R}^n$ such that $X \cap [U(a) \times \mathbb{R}^m]$ is described by $\mathcal{M}(U(a))[y_1, \ldots, y_m]$ where the m-tuple $(y_1, \ldots, y_m)$ denotes the last m components of points $x \in X$.

Figure 22.1 below illustrates part (ii) of this definition. We note the following useful properties of the introduced concepts.

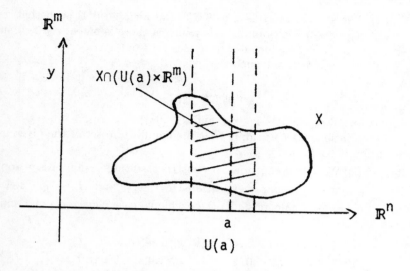

Fig. 22.1: Semi-Algebraic Analytical Set

Proposition 22.4. *(i) The classes of semi-algebraic, semi-analytical, or $\mathbb{R}^m$-semi-algebraic-analytical sets are closed under finite intersection, union, and difference.*

(ii) If $\alpha_1, \ldots, \alpha_k, \beta_1, \ldots, \beta_l, \gamma_1, \ldots, \gamma_r$ are functions from $\mathcal{M}(U)$ or from $\mathcal{M}(U)[y_1, \ldots, y_m]$, then

$$\{z \in U | \alpha_i(x) > 0, \beta_j(z) = 0, \gamma_q(z) \neq 0$$
$$\text{for all } i = 1, \ldots, k, \ j = 1, \ldots, l, \text{ and } q = 1, \ldots, r\}$$

is described by $\mathcal{M}(U)$, and

$$\{z \in U \times \mathbb{R}^m | \alpha_i(z) > 0, \beta_j(z) = 0, \gamma_q(z) \neq 0$$
$$\text{for all } i = 1, \ldots, k, \ j = 1, \ldots, l,$$
$$\text{and } q = 1, \ldots, r\}$$

is described by $\mathcal{M}(U)[y_1, \ldots, y_m]$.

The proof is immediate from the definitions. Note furthermore that the analogous statement clearly is also true for $\mathbb{R}[x_1, \ldots, x_n]$ in place of $\mathcal{M}(U)$. After these preparations we can start the main line of our proof of Theorem 11.8. First, let us assume without loss of generality that $\epsilon < 1$.

Step 1: In the first step we will show that there is a $\beta_0 > 0$ such that the equilibrium set of the given evolution even is contained in $W = (\Delta^{n-1} \times [0, \beta_0]) \times [0, 1]$. From the properties of an evolution and Proposition 6.2 follows immediately:

(i) there is a $\overline{w} \in \mathbb{R}_+^n$ such that $\zeta_s(p, r) \geq -\overline{w} + (1, \ldots, 1)$ on $\Delta^{n-1} \times \mathbb{R}_+$ for all $s \in [0, 1]$.

(ii) The set $\{x|$ there is a $y \in \mathbb{R}^m_+$ and an $s \in [0,1]$ such that $x = A_s y \geq -\overline{w}\}$ is bounded, that means there is a real $\alpha_{\overline{w}} > 0$ such that $||x|| \leq \alpha_{\overline{w}}$ for all x from this set, and

(iii) there is a $\beta_0 > 0$ such that

$$||\zeta_s(p,r)|| > \alpha_{\overline{w}} + 1$$

for all $p \in \Delta^{n-1}$, $r \geq \beta_0$, and $s \in [0,1]$.

Observations (i)–(iii) imply that the equilibrium set of the given evolution is contained in W.

Step 2: (i) We are going now to construct suitable ϵ-uniformly approximating one-parametrizations $(\hat{\zeta}_s)_{s \in [0,1]}$ and $(\hat{t}_s)_{s \in [0,1]}$. Let be $\epsilon' := \frac{\epsilon}{4n}$. Analogously to the exchange case choose n polynomials $\overline{\zeta}_i$,

$i = 1, \ldots, n$, on $\mathbb{R}^n \times \mathbb{R} \times \mathbb{R}$ with $|\overline{\zeta}_i(p,r,s) - \zeta_{i_s}(p,r)| < \epsilon'$ on $\Delta^{n-1} \times [0, 2\beta_0] \times [0,1] =: W'$ for all i. Clearly, $\overline{\zeta} := (\overline{\zeta}_1, \ldots, \overline{\zeta}_n)$ satisfies

$$||\overline{\zeta} - \zeta|| < \sqrt{n}\epsilon'$$

on W'.

Define further

$$\overline{t}_s(p,r) := r - p \cdot \overline{\zeta}(p,r,s).$$

Then

$$|\overline{t}_s(p,r) - t_s(p,r)| \leq ||p|| \cdot ||\overline{\zeta}(p,r,s) - \zeta_s(p,r)|| \leq \sqrt{n}\epsilon'$$

on W' because $||p|| \leq 1$. Now put

$$\overline{\overline{t}}_s(p,r) := \overline{t}_s(p,r) + \sqrt{n}\epsilon',$$
$$\overline{\overline{\zeta}}_{i_s}(p,r) := \overline{\zeta}_i(p,r,s) - \sqrt{n}\epsilon', \; i = 1, \ldots, n,$$
$$\text{and } \overline{\overline{\zeta}}_s := (\overline{\overline{\zeta}}_{1_s}, \ldots, \overline{\overline{\zeta}}_{n_s}).$$

By construction $\overline{\overline{\zeta}}_s$ and $\overline{\overline{t}}_s$ satisfy Walras' law on W' for any fixed s, and furthermore $\overline{\overline{t}} \geq 0$ on W'. Moreover, on W' we have

$$||\overline{\overline{\zeta}} - \zeta|| \leq ||\overline{\overline{\zeta}} - \overline{\zeta}|| + ||\overline{\zeta} - \zeta|| < n\epsilon' + \sqrt{n}\epsilon' < 2n\epsilon' = \epsilon,$$

and $||\overline{\overline{t}} - t|| < 2\sqrt{n}\epsilon' \leq \epsilon$. Now let us choose a gluing C^1-function

$$v : \mathbb{R} \longrightarrow [0,1]$$

with $v|_{[0,\beta_0]} \equiv 0$ and $r|_{[2\beta_0,\infty[} \equiv 1$, and define

$$\hat{\zeta}_s(p,r) := v(r)\zeta_s(p,r) + (1 - v(r))\bar{\bar{\zeta}}_s(p,r)$$
$$\text{and} \quad \hat{t}_s(p,r) := r - p \cdot \hat{\zeta}_s(p,r)$$
$$= v(r)t_s(p,r) + (1 - v(r))\bar{\bar{t}}_s(p,r).$$

Thus, the functions $\hat{\zeta}_s$ and $\hat{t}_s$ are C^1-functions for any s, and on $\Delta^{n-1} \times \mathbb{R}_+ \times [0,1]$ we have

$$||\hat{\zeta} - \zeta|| \le (1 - v(r))||\hat{\zeta} - \zeta|| < \epsilon,$$

and

$$||\hat{t} - t|| < \epsilon, \hat{t} \ge 0.$$

Furthermore, $\hat{t}$ is homogeneous of degree one as desired.

(ii) Now we set about constructing the ϵ-uniformly approximating one-parametrizations $(A_s)_{s\in[0,1]}$ and $(A_s^*)_{s\in[0,1]}$. First consider the continuous one-parametrization $(\sigma_s)_{s\in[0,1]}$ where σ_s is the $n \times m$ matrix of the tax rates at state s. Or more detailed

$$a_{ij_s}^* := a_{ij_s} - \sigma_{ij_s}|a_{ij_s}|.$$

Clearly, for all real $\eta > 0$ we can choose a finitely piecewise linear path $(\hat{\sigma}_s)_{s\in[0,1]}$ in $M(n \times m, \mathbb{R})$ with $\hat{\sigma}_{ij_s} \in [0,1]$ for all i, j, s such that

$$||\hat{\sigma}_s - \sigma_s|| < \eta \text{ for all } s \in [0,1].$$

(Notice that $(\sigma_s)_{s\in[0,1]}$ particularly is a continuous path in $[0,1]^{nm} \subset \mathbb{R}^{nm}$.)

Now let us come back to the given one-parametrization $(A_s)_{s\in[0,1]}$ of production matrices and let us begin with the following

Observation *From the properties of evolutions as we have constructed them in Part I and from Proposition 6.2 we know that $(A_s)_{s\in[0,1]}$ is not only a continuous path in the space of all $m \times n$-matrices with real coefficients, $M(m \times n, \mathbb{R})$, but even lies in the subspace*

$$M(\overline{w}, \alpha_{\overline{w}}) := \{A \in M(n \times m, \mathbb{R}) | A(\mathbb{R}_+^m) \cap \mathbb{R}_+^n = \{0^n\},$$
$$Ae^{m-j+1} = -1 \text{ for } j = 1, \ldots, n, \text{ and}$$
$$\text{whenever } x \in A(\mathbb{R}_+^m) \text{ with } x \ge -\overline{w}, \text{ then}$$
$$||x|| \le \alpha_{\overline{w}}\}$$

for $\overline{w} \in \mathbb{R}_+^n$ and $\alpha_{\overline{w}} > 0$ as in Step 1 above. (Remember that $e^i = (0, \ldots, 0, 1, 0, \ldots, 0) \in \mathbb{R}^m$ denotes the i-th unit vector.)

The following nice property of $M(\overline{w}, \alpha_{\overline{w}})$ provides the key for the desired approximation of the one-parametrization $(A_s)_{s\in[0,1]}$:

Lemma 22.5. *For any* $w \in \mathbb{R}^n_+$ *and any real* $\alpha > 0$ *the set* $M(w, \alpha) \subset M(n \times m, \mathbb{R}^n) \cong \mathbb{R}^{n \cdot m}$ *is semi-algebraic.*

Let us postpone the proof of Lemma 22.5 a little bit since now the moment has come to introduce a well-known mathematical result which will be of eminent significance for our proof.

Theorem 22.6. (Tarski-Seidenberg-Lojasiewicz)
Let be $X \subset \mathbb{R}^n \times \mathbb{R}^m$ *and* $p_2 : \mathbb{R}^n \times \mathbb{R}^m \longrightarrow \mathbb{R}^m$ *the canonical projection* $p_2(x, y) = y$. *Then*

(i) X *is semi-algebraic* $\Rightarrow p_2(X)$ *is semi-algebraic*
(ii) X *is* $\mathbb{R}^m$*-semi-algebraic-analytical* $\Rightarrow p_2(X)$ *is semi-analytical.*

Remark 22.7. In (ii) it would not suffice to require that X is semi-analytical (there are counterexamples).

The proof of Theorem 22.6 can be found in the articles by Lojasiewicz in the reference list at the end of the book.

Armed with the very strong result of Theorem 22.6 the proof of Lemma 22.5 is straightforward.

Proof of Lemma 22.5   Denote by $p : \mathbb{R}^m \times M(n \times m, \mathbb{R}) \longrightarrow M(n \times m, \mathbb{R})$ the canonical projection. Define

$$Z_1 \quad := \{(y, A)|y \in \mathbb{R}^m_+, A \in M(n \times m, \mathbb{R}), Ay \geq 0, Ay \neq 0\},$$

$$Z_2 \quad := \{(y, A)|y \in \mathbb{R}^m_+, A \in M(n \times m, \mathbb{R}), Ay \geq -w, \|Ay\| > \alpha\},$$

$$Z_3 \quad := \{(y, A)|y \in \mathbb{R}^m_+, A \in M(n \times m, \mathbb{R}), Ae^m + 1 = 0\},$$

$$Z_4 \quad := \{(y, A)|y \in \mathbb{R}^m_+, A \in M(n \times m, \mathbb{R}), Ae^{m-1} + 1 = 0\},$$

$$\vdots \qquad\qquad \vdots$$

$$Z_{n+2} := \{(y, A)|y \in \mathbb{R}^m_+, A \in M(n \times m, \mathbb{R}), Ae^{m-n+1} + 1 = 0\},$$

Note that all sets $M(n \times m, \mathbb{R}) \cong \mathbb{R}^{nm}$ and $Z_1, \ldots, Z_{n+2}$ are semi-algebraic. Thus by Theorem 1 also the sets $p(Z_1), \ldots, p(Z_{n+2})$ are semi-algebraic. Evidently,

$$M(w, \alpha_w) = (M(n \times m, \mathbb{R}) \backslash p(Z_1)) \cap (M(n \times m, \mathbb{R}) \backslash p(Z_2))$$
$$\cap\, p(Z_3) \cap \ldots \cap\, p(Z_{n+2}).$$

Since finite differences and intersections of semi-algebraic sets again are semi-algebraic, Lemma 22.5 is proven.

Now we can prove the existence of uniformly approximating paths of production matrices which even are finitely piecewise analytical paths in $M(n \times m, \mathbb{R}) \cong \mathbb{R}^{n \cdot m}$.

Proposition 22.8. *For any given continuous path (one-parametrization)* $(A_s)_{s\in[0,1]} : [0,1] \longrightarrow M(w, \alpha_w)$ *with arbitrary* $w \in \mathbb{R}^n_+$ *and* $\alpha_w > 0$ *and any* $\eta > 0$ *there is a finitely piecewise analytical path*

$$(\hat{A}_s)_{s\in[0,1]} : [0,1] \longrightarrow M(w, \alpha_w)$$
$$with \ \forall_{s\in[0,1]} ||A_s - \hat{A}_s|| < \eta.$$

Proof.   Let us begin with the following observation: for any fixed $s \in [0,1]$ there is an open neighborhood $U(A_s)$ in $M(n \times m, \mathbb{R})$ such that $U(A_s) \cap M(w, \alpha_w)$ is piecewise analytically path connected. This follows directly from Theorem 11.7 (see the proof of Proposition 11.6; replace $\mathbb{R}^n$ in Theorem 11.7 by $\mathbb{R}^{n \cdot m}$, A by $M(w, \alpha_w)$, and y by A_s).
Without loss of generality let us assume that $U(A_s)$ is a ball with center A_s and radius $\eta/2$. Now let us choose a finite covering $U_1, \ldots, U_k$ of the arc $\bigcup_{s\in[0,1]} A_s = (A_s[0,1])_{s\in[0,1]}$ consisting of such balls. Then clearly there exists a finite subdivision $0 = s_0 < s_1 < \ldots < s_l = 1$ and a mapping $\vartheta : \{1, \ldots, l\} \longrightarrow \{1, \ldots, k\}$ such that

$$\bigcup_{s\in[s_{i-1}, s_i]} A_s \subset U_{\vartheta(i)}.$$

According to the observation above we choose now l finitely piecewise analytical paths

$$a_i : [s_{i-1}, s_i] \longrightarrow U_{\vartheta(i)} \cap M(w, \alpha_w)$$
$$with \ a_i(s_{i-1}) = A_{s_{i-1}} \ and$$
$$a_i(s_i) = A_{s_i} \ for \ \ i = 1, \ldots, l.$$

By construction, the composition of the paths a_i yields a finitely piecewise analytical path

$$(\hat{A}_s)_{s\in[0,1]} : [0,1] \longrightarrow M(w, \alpha_w)$$

with the desired approximation property.
Putting now pieces together we see that choosing η small enough for the approximating paths of matrices $(\hat{A}_s))_{s\in[0,1]}$ and $(\hat{\sigma}_s)_{s\in[0,1]}$ we obtain two paths $(\hat{A}_s)_{s\in[0,1]}$ and $(\hat{A}_s^*)_{s\in[0,1]} = ((a_{ij_s} - \sigma_{ij_s} |a_{ij_s}|)_{ij_s\in[0,1]}$ with the desired properties. (Notice that $|\varphi| : [0,1] \longrightarrow \mathbb{R}_+$ is a finitely piecewise analytical path whenever $\varphi : [0,1] \longrightarrow \mathbb{R}$ is. This follows from the fact that a finitely piecewise analytical path has only finitely many zeroes (see Proposition 11.1).) Furthermore, the last statement of Theorem 22.6 is evident.

(iii) Finally, we have to convince ourselves that the equilibrium set of our constructed ϵ-uniformly approximating evolution $(\hat{\zeta}_s, \hat{t}_s, \hat{A}_s, \hat{A}^*_s)_{s \in [0,1]}$ also lies in the compact set $W = (\Delta^{n-1} \times [0, \beta_0]) \times [0, 1]$. Nevertheless, this is straightforward.

Since $\epsilon < 1$ we have by construction $\hat{\zeta}_s \geq -\overline{w}$ for any $s \in [0, 1]$ and $||(\hat{\zeta})s(p, r)|| \geq ||\zeta_s(p, r)|| - ||\zeta_s(p, r) - \hat{\zeta}_s(p, r)|| > (\alpha_{\overline{w}} + 1) - \epsilon > \alpha_{\overline{w}}$ for all $(p, r, s) \in \Delta^{n-1} \times [\beta_0, \infty[\times [0, 1]$. From step 1 above and Proposition 22.8 we know that $(\hat{A}_s)_{s \in [0,1]} : [0, 1] \longrightarrow M(\overline{w}, \alpha_{\overline{w}})$. Thus it follows from the equilibrium condition (E.2) that no equilibria of the approximating evolution can lie outside of W.

Step 3: In the last step we will demonstrate that the equilibrium set of the constructed approximating evolution has finitely piecewise analytical equilibrium paths. Our procedure will be to show that the equilibrium set is semi-analytical which in turn makes it amenable to a generalization of Proposition 11.6 to semi-analytical sets.

(i) For checking semi-analyticity of the equilibrium set of $(\hat{\zeta}_s, \hat{t}_s, \hat{A}_s, \hat{A}^*_s)_{s \in [0,1]}$ we first choose a finite subdivision $s_0 = 0 < s_1 < \ldots < s_b = 1$ of $[0, 1]$ such that for all $i = 1, \ldots, b$ the restrictions $(\hat{A}_s)_{s \in [s_{i-1}, s_i]}$ and $(A^*_s)_{s \in [s_{i-1}, s_i]}$ are analytical paths. Our next task will be to provide a certain $\mathbb{R}^m$-semi-algebraic-analytical subset $Z \subset \mathbb{R}^n \times \mathbb{R} \times [0, 1] \times \mathbb{R}^m$ which precisely projects on the equilibrium set under consideration. Let be

$$Z := \bigcup_{i=1}^{b} Z_i \text{ with}$$
$$Z_i := \{((p, r, s), y) \in \mathbb{R}^n \times \mathbb{R} \times [0, 1] \times \mathbb{R}^m | p \in \Delta^{n-1},$$
$$r \in [0, \beta_0], s \in [s_{i-1}, s_i], \ y \geq 0, p'\hat{A}^*_s \leq 0,$$
$$\hat{\zeta}_s(p, r) - \hat{A}_s y = 0, p'\hat{A}^*_s y = 0\} \text{ for } i = 1, \ldots, b.$$

It is not hard to see that all sets Z_i, and consequently also Z, are $\mathbb{R}^m$-semi-algebraic-analytical: one just has to observe that $(\hat{\zeta}_s)_{s \in [0,1]}$ equals $(\overline{\overline{\zeta}}_s)_{s \in [0,1]}$ on the relevant area $W = (\Delta^{n-1} \times [0, \beta_0]) \times [0, 1]$, and $(\overline{\overline{\zeta}}_s)_{s \in [0,1]}$ is by construction a *polynomial* on $\mathbb{R}^n \times \mathbb{R} \times \mathbb{R}$. As to the piecewise analytical paths $(\hat{A}_s)_{s \in [0,1]}$ and $(\hat{A}^*_s)_{s \in [0,1]}$ notice that analytical pahts are by definition restrictions of analytical functions which are defined on *open neighborhoods*. Keeping these facts in mind it is easy to convince oneself that any set Z_i actually is $\mathbb{R}^m$-semi-algebraic-analytical.

(ii) Now, by the Tarski-Seidenberg-Lojasiewicz-Theorem the projection $p(Z)$ is also semi-analytical where p is the canonical projection

$$p : (\mathbb{R}^n \times \mathbb{R} \times [0, 1]) \times \mathbb{R}^m \longrightarrow \mathbb{R}^n \times \mathbb{R} \times [0, 1].$$

But from the construction of the Z_i follows directly that $p(Z)$ equals the equilibrium set of the given evolution $(\hat{\zeta}_s, \hat{t}_s, \hat{A}_s, \hat{A}_s^*)_{s \in [0,1]}$.

(iii) Now, the last gap is filled in by the addressed generalization of Proposition 11.6.

Proposition 22.9. *Let $A \subset \mathbb{R}^n$ be semi-analytical. Then for any $y \in A$ there is an arbitrarily small open neighborhood $U(y)$ of y in $\mathbb{R}^n$ such that $U(y) \cap A$ is piecewise analytically path connected. Moreover, the connected components of A are even piecewise analytically path connected.*

The proof is completely analogous to the proof of Proposition 11.6.
Due to Proposition 22.9 joining equilibrium components of $(\hat{\zeta}_s, \hat{t}_s, \hat{A}_s, \hat{A}_s^*)_{s \in [0,1]}$ are piecewise analytically path connected, and thus there is at least one nicely behaved equilibrium path for $(\hat{\zeta}_s, \hat{t}_s, \hat{A}_s, \hat{A}_s^*)_{s \in [0,1]}$ by Theorem 10.2.

(II) We are still left with pointing out the necessary modifications of the presented proof for the generalized basic model with production, taxes, and subsidies from Section 6.2. In summary there are four differences to the model with production and taxes:

(1) the approximating one-parametrization $(\hat{t}_s)_{s \in [0,1]}$ must obey the additional condition $\hat{t}_s(p, r, 0) > 0$ for all $(p, r) \in \Delta^{n-1} \times \mathbb{R}_+$.

(2) there is another path of matrices $(A_s^{**})_{s \in [0,1]}$ with $a_{ij_s}^{**} = \chi_{ij_s} |a_{ij_s}|$, $\chi_{ij_s} \in [0, 1]$.

(3) the equilibrium conditions (E.1) and (E.3) are enlarged by a new summand γA_s^{**}.

(4) equilibria are now triples $(p^0, 0, \gamma^0)$ from $\Delta^{n-1} \times \mathbb{R}_+ \times [0, 1]$, or triples $(p^0, r^0, 1)$ from $\Delta^{n-1} \times \mathbb{R}_+ \times [0, 1]$.

Let us now see how to suitably adapt the constructions from (I) to meet these new requirements.

(1) There is a minimal value $\delta > 0$ of $t_s(p, r, 0) = r - p \cdot \zeta_s(p, r, 0)$ on the compact set $(\Delta^{n-1} \times [0, 2\beta_0] \times \{0\}) \times [0, 1]$. Now for the approximation the polynomial $\bar{\zeta}$ on $\mathbb{R}^n \times \mathbb{R} \times \mathbb{R} \times \mathbb{R}$ must be chosen so that it ϵ''-approximates $(\zeta_s)_{s \in [0,1]} |_{\Delta^{n-1} \times [0, 2\beta_0] \times \{0\}}$ with $\epsilon'' := \min(\delta/n, \frac{\epsilon}{4n})$. This is sufficient to ensure that $\hat{t}_s(p, r, 0) := v(r) t_s(p, r, 0) + (1 - v(r)) \bar{\bar{t}}_s(p, r, 0) > 0$ for all $s \in [0, 1]$ and $(p, r) \in \Delta^{n-1} \times \mathbb{R}_+$.

(2) There is no difficulty at all also to apply the method provided in step 2, (ii), above for achieving a piecewise analytical ϵ-approximating path $(\hat{A}_s^*)_{s \in [0,1]}$ for $(A_s^{**})_{s \in [0,1]}$.

(3),(4) Finally, it is also straightforward to incorporate the additional summand γA_s^{**} and the new domain of equilibria into the constructions of step 3 above. We just have to redefine $Z := \bigcup_{i=1}^b (Z_i^1 \bigcup Z_i^2)$ where

$$Z_i^1 := \{((p, r, \gamma, s), y) \in \mathbb{R}^n \times \mathbb{R} \times [0, 1] \times [0, 1] \times \mathbb{R}^m |$$

$$p \in \Delta^{n-1}, r \in [0, \beta_0], \gamma = 1, s \in [s_{i-1}, s_i], y \geq 0,$$
$$p'(\hat{A}_s^* + \hat{A}_s^{**}) \leq 0, \hat{\zeta}_s(p, r, 1) - \hat{A}_s y = 0,$$
$$p'(\hat{A}_s^* + \hat{A}_s^{**})y = 0\} \text{ for } i = 1, \ldots, b,$$

and

$$Z_i^2 := \{((p, r, \gamma, s), y) \in \mathbb{R}^n \times \mathbb{R} \times [0, 1] \times [0, 1] \times \mathbb{R}^m |$$
$$p \in \Delta^{n-1}, r = 0, \gamma \in [0, 1], s \in [s_{i-1}, s_i], y \geq 0,$$
$$p'(\hat{A}_s^* + \gamma \hat{A}_s^{**}) \leq 0, \hat{\zeta}_s(p, r, \gamma) - \hat{A}_s y = 0,$$
$$p'(\hat{A}_s^* + \gamma \hat{A}_s^{**})y = 0\} \quad \text{for} \quad i = 1, \ldots, b.$$

This completes our proof of Theorem 11.8.

$\square$

Appendix B

Proof of Theorem 13.1

Correspondingly to Mas-Colell's proof (1977, Section 3) the proof of our Theorem 13.1 will consist of two parts. In the *first part* we will provide a continuous one-parametrization $(f_s)_{s \in [0,1]}$ of market excess demand functions whose equilibrium set $\bigcup_{s \in [0,1]} (f_s^{-1}(0) \times \{s\})$ equals the prescribed set K. In general, however, there are no additional properties, like for instance differentiability, of the obtained one-parametrization $(f_s)_{s \in [0,1]}$. Our construction relies on a result from algebraic topological fixed point theory by H. Schirmer (1983). In the *second part* we will check that Mas-Colell's decomposing constructions (1977) in fact provide all necessary tools to extend the static context to a continuous one-parametrization of exchange economies $(E_s)_{s \in [0,1]}$ with the desired properties. Roughly spoken we will proceed by demonstrating that our obtained continuous one-parametrization $(f_s)_{s \in [0,1]}$ of market excess demand functions induces *continuous* one-parametrizations of the relevant constructions by Mas-Colell.

(I) At the heart of the first part of the proof stands the following intuitive result from algebraic topological fiberwise fixed point theory:

Proposition 23.1. (H. Schirmer 1983): *Let P be a compact and connected polyhedron without local cut points, i.e. without points p which have a connected neighbourhood N in P so that $N \setminus \{p\}$ is not connected. Let $K \subset P \times [0,1]$ be a closed set which contains a compact connected subspace $\widetilde{C}$ joining $P \times \{0\}$ and $P \times \{1\}$. Then there exists a continuous one-parametrization*

$$H : P \times [0,1] \to P$$

with

$$Fix\, H := \{(p,t) \in P \times [0,1] \mid H(p,t) = p\} = K.$$

This is proven by Schirmer (1983, Section 3).

Clearly, for $l \geq 3$ the closed positive orthant of the sphere S_+^{l-1} is a polyhedron without local cut points, and consequently qualifies for P in Proposition 23.1. On the contrary, S_+^1 obviously has a continuum of local cut points, namely S_{++}^1. Nevertheless, for $P = S_+^1$ we can prove Proposition 23.1 directly:

Proposition 23.2. *Let $K \subset S_+^1 \times [0,1]$ be a closed set which contains a compact connected subspace $\widetilde{C}$ joining $S_+^1 \times \{0\}$ and $S_+^1 \times \{1\}$. Then there exists a continuous one-parametrization*

$$H : S_+^1 \times [0,1] \rightarrow S_+^1$$

with

$$Fix\, H := \{(p,t) \in S_+^1 \times [0,1] | H(p,t) = p\} = K.$$

Proof. Map $S_+^1 \subset \mathbb{R}_+^2$ by the canonical homeomorphism onto the interval $[0,1]$. Thus $S_+^1 \times [0,1] \approx [0,1] \times [0,1] \subset \mathbb{R}^2$. Consider the complement K^c of K in $[0,1] \times [0,1]$. K^c is relatively open. Define

$$K^c \supset K^1 := \{(p,t) \in K^c | \text{there is a point } (1,s) \in K^c \text{ so that } (p,t) \text{ lies}$$
$$\text{in the relatively open connected component of } K^c$$
$$\text{which contains } (1,s)\}.$$

In other words,

$$K^1 = \bigcup L \,, \text{where any } L \subset K^c$$

is a relatively open component of K^c with $L \cap (\{1\} \times [0,1]) \neq \emptyset$.
Clearly, K^1 is open, and possibly it is empty (see Figure 23.1 for an illustration of a non-empty set).
Now we can compose a homotopy H with the desired properties.

$$H : [0,1] \times [0,1] \longrightarrow \mathbb{R}$$
$$(p,t) \mapsto \begin{cases} p - \text{dist}[(p,t), K] & \text{if } (p,t) \in K^1 \\ p + \text{dist}[(p,t), K] & \text{if } (p,t) \notin K^1. \end{cases}$$

Let us check the required properties of H: clearly, $Fix\, H = K$. So it is left to verify that (a) every t-state mapping $H_t : [0,1] \rightarrow \mathbb{R}$, $p \mapsto H(p,t)$, actually is a *self-mapping* of $[0,1]$, and (b) that H is continuous.
Both properties are obvious if K^1 is empty. Let us therefore assume that K^1 is not empty.
(a) To show (a) we have to ensure that

$$0 \leq p - \text{dist}[(p,t), K] \qquad \text{for } (p,t) \in K^1$$
$$\text{and } p + \text{dist}[(p,t), K] \leq 1 \quad \text{for } (p,t) \notin K^1.$$

But for every point $(p,t) \in K^1$ there are points $(p',t) \in K$ which are 'left' of (p,t), i.e. $p' < p$. Otherwise $(p,t) \in K^1 \subset K^c$ could be

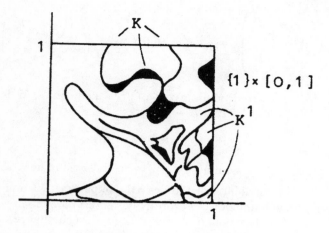

Fig. 23.1: Auxiliary Set K_1

connected with $(0, t)$ in K^c, and this would contradict the assumed existence of a joining connected component $\widetilde{C}$ in K. This proves the first inequality.

If, on the other hand, $(p, t) \in (K^1)^c$, then either $p = 1$ and $(p, t) \in K$, or there are points $(p'', t) \in K$ 'right' of (p, t), i.e. $p'' > p$. Consequently, $p + \text{dist}[(p, t), K] \leq 1$ for $(p, t) \in (K^1)^c$.

(b) We have to show that H is continuous. Clearly, $H|_{K^1}$ and $H|_{int(K^1)^c}$ are continuous because K^1 is open. Obviously, the common boundary of K^1 and $(K^1)^c$ in $]0, 1[\times]0, 1[$ is contained in K, i.e.

$$(\bar{K}^1 \setminus K^1) \cap]0, 1[\times]0, 1[=$$
$$[(K^1)^c \setminus int(K^1)^c] \cap]0, 1[\times]0, 1[\subset K.$$

Since $\text{dist}[(p, t), K]$ is zero on K, this implies that H is continuous, and Proposition 23.2 is proven.

After these preparatory results we will continue the proof of the Theorem. Our task will be to appropriately modify the one-parametrization H near the boundary of S_+^{l-1} such that a one-parametrization of excess demand functions obtains.

Take the continuous perturbation

$$H : S_+^{l-1} \times [0, 1] \to S_+^{l-1}$$

provided by Propositions 23.1 and 23.2. Now consider the modified continuous perturbation

$$Z := (H - id) : S_+^{l-1} \times [0, 1] \to \mathbb{R}^l$$

$$(p, s) = \begin{pmatrix} p_1 \\ \vdots \\ p_l \\ s \end{pmatrix} \mapsto Z_s(p) = H_s(p) - p = \begin{pmatrix} H_{1_s}(p) - p_1 \\ \vdots \\ H_{l_s}(p) - p_l \end{pmatrix}.$$

Clearly, the set of fixed points $\text{Fix}\, H$ of H, which equals K, is identical with the set of zeroes of $Z = H - id$:

$$\text{Fix}\, H = (H - id)^{-1}(0).$$

Actually, $Z|_{S_{++}^{l-1} \times [0,1]}$ is almost – up to its behaviour near the boundary ∂S_+^{l-1} – our candidate for the desired one-parametrization $(f_s)_{s \in [0,1]}$ of excess demand functions. Accordingly, let us now modify Z over a neighbourhood of the boundary $\partial S_+^{l-1} \times [0,1]$ such that a continuous one-parametrization of market excess demand functions with all desired properties obtains. Obviously, there are many ways to do that. Here we propose the following one because of its intuitive appeal:
fix an $s \in [0,1]$ and consider the first $l-1$ component functions $Z_{1_s}, \ldots,$ Z_{l-1_s} of Z_s. Choose a positive δ such that $K \subset S_\delta^{l-1} \times [0,1]$ (recall that K is a compact subspace of $S_{++}^{l-1} \times [0,1]$), and restrict the $Z_{i_s}, i = 1, \ldots, l-1,$ to S_δ^{l-1}. Now take one of these component functions, say $Z_{1_s}|_{S_\delta^{l-1}}$, and replace it by the function of its absolute values

$$|Z_{1_s}|_{S_\delta^{l-1}}(p)|.$$

Now replace Z_{l_s} by the function $\widehat{Z}_{l_s}$ which is obtained from the Walras formula (the budget identity):

$$\text{for } p \in S_\delta^{l-1} \text{ define } \widehat{Z}_{l_s}(p) := -\frac{p_1}{p_l}|H_{1_s}(p)| - \sum_{i=2}^{l-1} \frac{p_i}{p_l} H_{1_s}(p).$$

For the following constructions we need some technical preparations: let us partition the neighbourhood $S_+^{l-1} \setminus S_\delta^{l-1}$ of the boundary ∂S_+^{l-1} into the three areas (note that S_ϑ^{l-1} is closed for any positive ϑ)

$$A := S_{\frac{2}{3}\delta}^{l-1} \setminus S_\delta^{l-1}$$

$$B := S_{\frac{\delta}{3}}^{l-1} \setminus S_{\frac{2}{3}\delta}^{l-1}$$

$$C := S_{++}^{l-1} \setminus S_{\frac{\delta}{3}}^{l-1}.$$

To simplify the geometrical representation we illustrate this partition on the unit simplex Δ^{l-1} (see Figure 23.2 below).

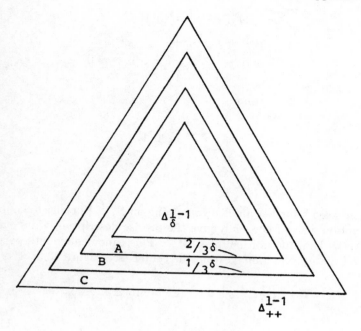

Fig. 23.2: Partition of the Unit Simplex

Now choose some arbitrary market excess demand function $\bar{\zeta} = \begin{pmatrix} \bar{\zeta}_1 \\ \vdots \\ \bar{\zeta}_l \end{pmatrix}$ with a *unique* equilibrium in the center $(\frac{1}{\sqrt{l}}, \ldots, \frac{1}{\sqrt{l}})$ of S_{++}^{l-1} and with the additional property that the zero sets of $\bar{\zeta}_2$ to $\bar{\zeta}_{l-1}$ have an empty intersection in $S_{++}^{l-1} \setminus S_\delta^{l-1}$, i.e.

$$\bigcap_{i=2}^{l-1} (\bar{\zeta}_i|_{S_{++}^{l-1} \setminus S_\delta^{l-1}})^{-1}(0) = \emptyset.$$

(Actually, there is a myriad of such market excess demand functions.) Choose two continuous gluing functions

$$rcl \ \alpha : S^{l-1} \rightarrow [0,1]$$
$$\text{with } \alpha \mid_{S_\delta^{l-1}} = 1$$
$$\text{and } \alpha \mid_{S_+^{l-1} \setminus S_{\frac{2}{3}\delta}^{l-1}} = \alpha \mid_{B \cup C} = 0$$

and

$$\beta : S^{l-1} \rightarrow [0,1]$$
$$\text{with } \beta \mid_{S_{\frac{2}{3}\delta}^{l-1}} = \beta \mid_{S_\delta^{l-1} \cup A} = 1$$
$$\text{and } \beta \mid_C = 0.$$

We still need a further auxiliary mapping on S_+^{l-1}. In order to simplify its description we will here only give the precise description of its canonical counterpart on the unit simplex Δ^{l-1}. This canonical counterpart is the natural *radial retraction* $\widetilde{\gamma}_\delta$ from the center (see Figure 23.3)

$$\left(\frac{1}{l}, \dots, \frac{1}{l} \right) \text{ of } \Delta^{l-1} \text{ onto } \Delta_\delta^{l-1}, \text{ i.e.}$$

$$\widetilde{\gamma}_\delta : \Delta^{l-1} \rightarrow \Delta_\delta^{l-1}$$
$$\text{with } \widetilde{\gamma}_\delta \mid_{\Delta_\delta^{l-1}} = id_{\Delta_\delta^{l-1}}$$
$$\text{and } \widetilde{\gamma}_\delta(\Delta^{l-1}) = \Delta_\delta^{l-1}.$$

Let us denote the canonical counterpart of $\widetilde{\gamma}_\delta$ on S_+^{l-1}, i.e. the radial retraction mapping from S_+^{l-1} onto S_δ^{l-1}, by γ_δ. After these preparations we appropriately extend our restricted candidate function

$$(|Z_{1_s}(p)|, Z_{2_s}(p), \dots, Z_{l-1_s}(p), \widehat{Z}_{l_s}(p)) : S_\delta^{l-1} \rightarrow \mathbb{R}^l$$

continuously from S_δ^{l-1} to S_{++}^{l-1} in three steps
(1) <u>extension over A:</u> let be $p \in A$.
  (i) Roughly speaking $|Z_{1_s}(p)|$ is extended by gluing it with the constant function $+1$. Formally

$$p \mapsto \alpha(p)|Z_{1_s}(\gamma_\delta(p))| + (1 - \alpha(p)) \cdot 1$$

  Note that the image values are positive for $p \in A$.
  (ii) $Z_{2_s}, \dots, Z_{l-1_s}$ are extended accordingly by gluing them with the corresponding component functions ζ_i, i.e.

$$p \mapsto \alpha(p)Z_{i_s}(\gamma_\delta(p)) + (1 - \alpha(p))\bar{\zeta}_i(p),$$

$$i = 2, \dots, l-1$$

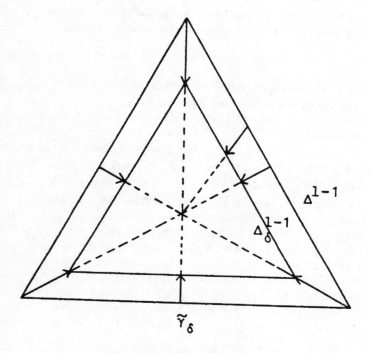

Fig. 23.3: Radial Retraction

(iii) $\widehat{Z}_{l_s}$ will be extended simultaneously over A and B in the next step.
(2) <u>extension over B:</u> let be $p \in B$.
  (i) glue the constant $+1$-function with $\bar{\zeta}_1$ by means of β, i.e.

$$p \mapsto \beta(p) + (1 - \beta(p))\bar{\zeta}_1(p)$$

  (ii) map

$$p \mapsto \bar{\zeta}_i(p)$$

  for $i = 2, \ldots, l - 1$.
  (iii) extend $\widehat{Z}_{l_s}(p)$ over $A \cup B$ according to the Walras formula.
(3) <u>extension over C:</u> for any $p \in C$ take $\bar{\zeta}_1(p), \ldots, \bar{\zeta}_l(p)$.
Clearly, the resulting function

$$f_s : S_{++}^{l-1} \to \mathbb{R}^l$$
$$p \mapsto (f_{1_s}(p), \ldots, f_{l_s}(p))$$

is a market excess demand function which has no zeroes in $S_{++}^{l-1} \setminus S_\delta^{l-1}$
and for which

$$f_s^{-1}(0) = K \cap S_{++}^{l-1} \times \{s\}.$$

Furthermore, from the construction of f_s follows immediately that we actually have achieved a *continuous one-parametrization*

$$(f_s)_{s\in[0,1]} : S_{++}^{l-1} \times [0,1] \to \mathbb{R}^l$$

of market excess demand functions with an equilibrium set $\bigcup_{s\in[0,1]} f_s^{-1}(0) \times \{s\}$ equal to the prescribed set K.

Thus we have completed the first part of our proof. Let us now proceed to the second part.

(II) Given any positive ε with $\delta/\varepsilon > \varepsilon > 0$ we will provide a continuous one-parametrization of exchange economies $(E_s)_{s\in[0,1]} = \left((\preceq_{i_s}, \omega_{i_s})_{i=1}^l\right)_{s\in[0,1]}$ with $\preceq_{i_s} \in \mathcal{P}_{mo}^0$ and $\omega_{i_s} \in \mathbb{R}_+^l$ such that for any $s \in [0,1]$ the excess demand function f_s equals the derived excess demand function of E_s on S_ε^{l-1} and, in addition, the equilibrium set of E_s equals the zero set of f_s. More precisely the latter means that $p \in S_{++}^{l-1}$ is a zero of f_s if and only if it is an equilibrium of E_s, i.e. a zero of its derived excess demand function. Our procedure will be to verify that the constructions in Mas-Colell (1977, Section 3, Proof of the Theorem) lead to a *continuous* one-parametrization $(E_s)_{s\in[0,1]}$ of exchange economies when the underlying market excess demand function is continuously perturbed.

Let us fix an arbitrary $s \in [0,1]$. Consider the s-state excess demand function

$$f_s : S_{++}^{l-1} \to \mathbb{R}^l$$

of the continuous one-parametrization $(f_s)_{s\in[0,1]}$ obtained at the end of part (I). Going through the relevant constructions by Mas-Colell we will see that most of them are independent of the very function f_s. Thus it will be left to us to ensure that the constructions which are dependent on f_s get continuously one-parametrized when f_s is continuously one-parametrized. Analogously to the proof by Mas-Colell we will proceed in *three main steps*. In the **first step** we will ensure that the continuous one-parametrization $(f_s)_{s\in[0,1]}$ induces *continuous one-parametrizations* $(f_{i_s}^*)_{s\in[0,1]}$, $i = 1, \ldots, l$, of the l *decomposing individual* excess demand functions $f_{i_s}^*$ which are generally provided by Mas-Colell on p. 125 in (1977). (To be more precise, the l individual excess demand functions decompose the given excess demand functions only on S_ε^{l-1}.) In the **second step** we will then verify that the induced continuous one-parametrization $(f_{i_s}^*)_{s\in[0,1]}$ of each individual excess demand function in turn induces a continuous one-parametrization of the monotone preference relation which is associated to it by Mas-Colell's construction on pp. 121–123 in (1977, Proof of Lemma 2). Finally in the **third step** we will verify that Debreu's (1974) construction of strict convexification also applies to our situation and leads to a continuous one-parametrization of exchange economies with the desired properties.

Step 1: The following preparatory result actually poses ourselves into Mas-Colell's set-up (cf. 1977, Section 3, (1)).

Lemma 23.3. *For the continuous one-parametrization*

$$(f_s)_{s\in[0,1]} : S_{++}^{l-1} \times [0,1] \to \mathbb{R}^l$$

of market excess demand functions obtained at the end of Part I there is a $\tau > 0$ such that for all $(p,s) \in (S_{++}^{l-1} \setminus S_\tau^{l-1}) \times [0,1]$ one has:

$$f_s(p) \left[e - (\textstyle\sum_{i=1}^l p_i)p \right] = \sum_{i=1}^l f_{i_s}(p) > 0.$$

Proof. Remember that by construction the s-state excess demand functions for all $s \in [0,1]$ are identical on $C = S_{++}^{l-1} \setminus S_{\delta/3}^{l-1}$. Thus Lemma 23.3 is an easy Corollary of Lemma 1 in Mas-Colell (1977).
Particularly, the prescribed set $K \subset S_{++}^{l-1} \times [0,1]$ is contained in $S_\tau^{l-1} \times [0,1]$. Let be $\varepsilon \in]0,\tau[$. Define l individual excess demand functions

$$f_{i_s}^* : S_{++}^{l-1} \to \mathbb{R}^l, \qquad i = 1,\dots,l,$$

by

$$f_{i_s}^*(p) := -[\eta(p) + (1 - \eta(p))\beta_{i_s}(p)] \ \mathrm{grad}\, v_i(p).$$

Furthermore, define $f_s^*(p) := \sum_{i=1}^l f_{i_s}^*(p)$. Now we have to explain the ingredients of $f_{i_s}^*(p)$. Let us start with those which are independent of f_{i_s}:

$$\eta : S_{++}^{l-1} \to [0,1]$$

is some C^2 gluing function with

$$\eta|_{S_{++}^{l-1} \setminus S_{\frac{\varepsilon}{2}}^{l-1}} = 1$$

and

$$\eta|_{S_\varepsilon^{l-1}} = 0.$$

$$v_i : S_{++}^{l-1} \to \mathbb{R}$$

is a fixed function with certain properties (see Mas-Colell (1977), pp. 121–125).
The only ingredients of $f_{i_s}^*(p)$ which depend on f_{i_s} are the coefficients $\beta_{i_s}(p) \in \mathbb{R}_+$. They are determined in the following way:
let Γ denote a closed convex cone in $\mathbb{R}_+^n$ with vertex 0 which is spanned by an l-tuple of linearly independent vectors $(a_1,\dots,a_l)$ of S_{++}^{l-1} such that

$$S_{\frac{\varepsilon}{2}}^{l-1} \subset int\,\Gamma \quad \text{and} \quad (\Gamma \cap S_{++}^{l-1}) \subset S_\mu^{l-1} \subset S_{++}^{l-1}$$

with an arbitrary $\mu \in]0, \frac{\varepsilon}{2}[$.
For any $r > 0$ there is a $\vartheta(r) > 0$ such that

$$\forall_{x \in B_r^l(0)} \ \forall_{p \in S_{\frac{\varepsilon}{2}}^{l-1}} \ (x + \vartheta(r)p) \in \Gamma$$

(recall that $S_{\frac{\varepsilon}{2}}^{l-1} \subset int\,\Gamma$).
Now pick an

$$\widetilde{r} > \max_{(p,s) \in S_\mu^{l-1} \times [0,1]} ||f_{i_s}(p)||.$$

Clearly, $\forall_{p \in S_{\varepsilon/2}^{l-1}} \forall_{s \in [0,1]} \ f_{i_s}(p) + \vartheta(\widetilde{r})p \in int\,\Gamma$. Since $a_1, \ldots, a_l$ are linearly independent one can write for $p \in S_{\frac{\varepsilon}{2}}^{l-1}$ and $s \in [0,1]$

$$f_{i_s}(p) + \vartheta(\widetilde{r})p = \sum_{i=1}^{l} \beta_{i_s}(p)a_i$$

with $\beta_{i_s}(p) > 0$

in a *unique* and *continuous* (in p and s) manner. From this follows immediately that for every $i, \ldots, l$ the one-parametrization $(f_{i_s}^*)_{s \in [0,1]}$ is also *continuous*.

Projecting now the vector $f_{i_s}(p) + \vartheta(\widetilde{r})p$ orthogonally on $T(p)$, i.e. on the normal hyperplane to p, one obtains from the definition of v_i the following equation (Mas-Colell, 1977, p. 125):

$$f_{i_s}^*(p) = \sum_{i=1}^{l} \beta_{i_s}(p)(-grad\,v_i(p)).$$

Furthermore, for all $i = 1, \ldots, l$ and $s \in [0,1]$, $f_{i_s}^*(p)$ and $f_s^*(p)$ are excess demand functions (see Mas-Colell, 1977, p. 125). By construction one has

$$\forall_{s \in [0,1]} f_s^*|_{S_\varepsilon^{l-1}} = f_s|_{S_\varepsilon^{l-1}},$$

and, moreover, there are no zeroes of f_s^* in $S_{++}^{l-1} \setminus S_\varepsilon^{l-1}$ (see Mas-Colell, 1977, p. 125, last paragraph). In other words, the l functions $f_{i_s}^*(p)$ decompose the given s-state market excess demand function f_s on S_ε^{l-1} in an equilibria preserving manner as desired.

Now we have to show that for each $f_{i_s}^*$ the rationalizing pair of an individual preference relation $\preceq_{i_s} \in \mathcal{P}_{mo}^0$ and an initial endowment bundle $\omega_{i_s} \in \mathbb{R}_+^l$ from Mas-Colell's construction (1977, proof of Lemma 2) is *continuously one-parametrized* when $(f_{i_s}^*)$ is continuously one-parametrized.

Step 2: Fix an $i \in \{1, \ldots, l\}$ and take the continuous one-parametrization

$$(f_{i_s}^*)_{s \in [0,1]} : S_{++}^{l-1} \times [0,1] \to \mathbb{R}^l$$

from Step 1. Fix also an $s \in [0,1]$. For $f_{i_s}^*$ Mas-Colell's constructions in (1977, proof of Lemma 2) provide an individual preference relation and an initial endowment bundle with all desired properties.

First we will recall this static construction, and then we will verify that it is continuously one-parametrized in dependence on $(f_{i_s}^*)_{s\in[0,1]}$. Analogously to Step 1, we only have to take care of the ingredients of this construction which are dependent on the very function $f_{i_s}^*$.

Let us start with introducing the ingredients which are *independent* of $f_{i_s}^*$: recall recall that $f_{i_s}^*(p) := -[\eta(p) + (1 - \eta(p))\beta_{i_s}(p)]\, grad\, v_i(p)$. v_i is a fixed function from S_{++}^{l-1} into $\mathbb{R}$ (see Mas-Colell, 1977, p. 121), and the real coefficient $[\eta(p) + (1 - \eta(p))\beta_{i_s}(p)]$ is positive and uniformly bounded away from 0. From the properties of v_i follows that v_i is minimized in at most one point. Hence there is a *unique* $\bar{p}^i \in S_{++}^{l-1}$ such that $f_{i_s}^*(\bar{p}^i) = 0$ (Mas-Colell, 1977, pp. 121–122). Thus, $\bar{p}^i$ is the only zero of *any* $f_{i_s}^*(p)$, $s \in [0, 1]$. Now $\mathbb{R}^l$ is partitioned into the following three regions:

$$\widehat{P} := \mathbb{R}_+^l \setminus \{0\},$$
$$A^1 := \{x \in \mathbb{R}^l | \exists_i x_i < 0 \text{ and } \bar{p}^i x > 0\}$$
$$B^1 := \{x \in \mathbb{R}^l | \bar{p}^i x \leq 0\}.$$

Actually, this partition is independent of $f_{i_s}^*$.

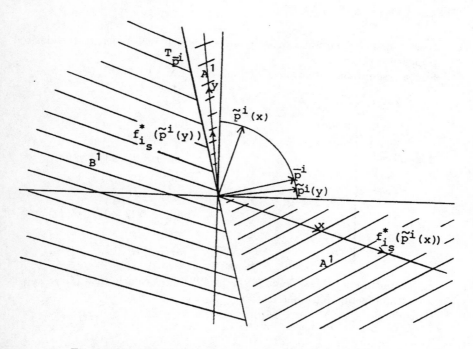

Fig. 23.4: Auxiliary Constructions I for Proving Lemma 23.3

Furthermore, choose some continuous function

$$u_i : S_{++}^{l-1} \to [0,1],$$

which is closely linked to v_i in the following way:

$$u_i(\bar{p}^i) = 0, \text{ and for all } p, q \in S_{++}^{l-1}$$
$$u_i(p\ ) > u_i(q) \Leftrightarrow v_i(p) > v_i(q)$$
$$\text{and } \ u_i(p\ ) = u_i(q) \Leftrightarrow v_i(p) = v_i(q).$$

In other words, u_i is a 'pre-image-preserving' and bounded modification of v_i.

We need a last preparatory observation: by the properties of v_i there is for every $x \in A^1$ a *unique* $\widetilde{p}^i(x) \in S_{++}^{l-1}$ such that x belongs to the *positive ray* spanned by $f_{i_s}^*(\widetilde{p}^i(x)) = -[\eta(\widetilde{p}^i(x)) + (1 - \eta(\widetilde{p}^i(x)))\beta_{i_s}(\widetilde{p}^i(x))] \cdot grad\, v_i(\widetilde{p}^i(x))$. In other words there is a unique $t_{i_s}(x) \in]0, \infty[$ such that $x = t_{i_s}(x) \cdot f_{i_s}^*(\widetilde{p}^i(x))$ (cf. Figure 23.4). The function $x \mapsto \widetilde{p}^i(x)$ is continuous (Mas-Colell, 1977, p. 122) and independent of s because it is only dependent on v_i. Obviously, the continuous one-parametrization $(f_{i_s})_{s\in[0,1]}$ also induces a continuous one-parametrization

$$(t_{i_s})_{s\in[0,1]} : A^1 \times [0,1] \to]0, \infty[$$
$$(x, s) \mapsto t_{i_s}(x)$$

Now a utility function can be provided which will generate the desired preference relation:

first a preliminary function is defined by

$$\xi_{i_s} : \mathbb{R}^l \to \mathbb{R}$$

$$x \mapsto \begin{cases} \min_i x_i & \text{for } x \in \widehat{P} \\ u(\widetilde{p}^i(x)) & \text{for } x \in A^1 \\ & \text{and } x = t_{i_s}(x)f_{i_s}^*(\widetilde{p}^i(x)) \\ & \text{with } 0 < t_{i_s}(x) \leq 1 \\ u(\widetilde{p}^i(x)) - ||x - f(\widetilde{p}^i(x))|| & \text{for } x \in A^1 \\ & \text{and } x = t_{i_s}(x)f_{i_s}^*(\widetilde{p}^i(x)) \\ & \text{with } t_{i_s}(x) > 1 \\ -||x|| & \text{for } x \in B^1 \end{cases}$$

ξ_{i_s} is continuous (Mas-Colell, 1977, p. 123), and clearly also $(\xi_{i_s})_{s\in[0,1]}$ is a continuous one-parametrization.

Now define

$$\widehat{\xi}_{i_s} : \mathbb{R}^l \to \mathbb{R}$$
$$x \mapsto \max\{\xi_{i_s}(y)|y \leq x\}$$

$\widehat{\xi}_{i_s}(x)$ is continuous, strictly monotone, i.e. $x > y$ implies $\widehat{\xi}_{i_s}(x) > \widehat{\xi}_{i_s}(y)$, and for any $p \in S_{++}^{l-1}$ and any $x \in \mathbb{R}^l$ one has:

$$px \leq 0 \ \text{ and } \ x \neq f^*_{i_s}(p) \Rightarrow \widehat{\xi}_{i_s}(x) < \widehat{\xi}_{i_s}(f^*_{i_s}(p))$$

(see Mas-Colell, 1977, p. 123). Clearly, $(\widehat{\xi}_{i_s})_{s\in[0,1]}$ is a *continuous one-para-metrization*.

The definition $x \succ'''''_{i_s} y \Leftrightarrow \widehat{\xi}_{i_s}(x) > \widehat{\xi}_{i_s}(y)$ provides a *continuous* and *monotone* preference relation $\preceq'''''_{i_s}$ on $\mathbb{R}^l$ (Mas-Colell, 1977, p. 123). Moreover it is clear that we have achieved a Hausdorff continuous one-parametrization of continuous and monotone preference relations $(\preceq'''''_{i_s})_{s\in[0,1]}$ on $\mathbb{R}^l$ for the i-th agent such that for any $s \in [0,1]$ the pair $(\preceq'''''_{i_s}, 0)$ generates $f^*_{i_s}(p)$ on the whole price space S^{l-1}_{++}.

Step 3: Now we are still left with the *following task*: to alter $\preceq'''''_{i_s}$ into a preference relation which is *strictly convex on* $\mathbb{R}^l_+$ and *generates* $f^*_{i_s}$ on S^{l-1}_{++} such that agent i's demand is *nonnegative* for all commodities and all prices.

We will proceed *in four steps*.

(1) *First*, we will *convexify* the preference relation $\preceq'''''_{i_s}$ by the following construction (c.f. Hildenbrand (1974), problem 1.1, 7): for $x \in \mathbb{R}^l$ $\psi'''''_{i_s}(x)$ denotes the closed upper contour set (the indifference or preference set) of x with respect to $\preceq'''''_{i_s}$. Define the *new convexified* closed upper contour set of x by

$$\psi''''_{i_s}(x) := \bigcap_{\substack{y\in\mathbb{R}^l \\ x\in \mathrm{co}\psi'''''_{i_s}(y)}} \mathrm{co}\,\psi'''''_{i_s}(y).$$

Clearly, the associated new preference relation $\preceq''''_{i_s}$ is continuous, monotone, and convex. Moreover, $\preceq''''_{i_s}$ generates $f^*_{i_s}$ on S^{l-1}_{++}, i.e., for any $p \in S^{l-1}_{++}$ and any $x \in \mathbb{R}^l$, $px \leq 0$ and $x \neq f^*_{i_s}(p)$ implies $x \prec''''_{i_s} f^*_{i_s}(p)$. This is an immediate consequence of the obvious relationship

$$\psi''''_{i_s}(x) \subset \mathrm{co}\psi'''''_{i_s}(x) \ \text{ for all } x \in \mathbb{R}^l.$$

This is true because in our situation the convexification process does not add any further best elements to any budget set. (Particularly, this is also a direct consequence of the more general result

$$\mathrm{co}\varphi(\preceq''''', w, p) = \varphi(\preceq'''', w, p)$$

for every $(w,p) \in \mathbb{R}_+ \times \mathbb{R}^l$ (Hildenbrand, 1974, p. 95) where $\varphi(\preceq, w, p)$ denotes the *demand set*

$$\{x \in \mathbb{R}^l | px \leq w \text{ and } py \leq w \text{ for } y \in \mathbb{R}^l \text{ implies } y \preceq x\}.)$$

Furthermore, from the definition it is clear that $(\psi''''_{i_s}(x))_{s\in[0,1]}$ is a *Hausdorff continuous* one-parametrization for any $x \in \mathbb{R}^l$.

Actually, for our further constructions we will only use the upper contour sets which are contained in (are "equal or better than") $\psi''''_{i_s}(0)$.

(2) Now we will apply Debreu's construction of *strict convexification* (1974, pp. 19-20) to $\preceq_{i_s}'''$ (cf. Mas-Colell, 1977, p. 122, proof of Lemma 2). It will turn out that also the strictly convexified preference relation is continuously one-parametrized in dependence on the continuous one-parametrization $(f_{i_s}^*)_{s\in[0,1]}$.

We have to show that Debreu's construction (1974, from the 3rd paragraph of p. 19 to p. 20) really applies to the situation of $\preceq_{i_s}'''$. We will proceed by first recalling the main line of Debreu's construction and then ensuring that our situation with $\preceq_{i_s}'''$ is essentially the same as Debreu's situation.

Debreu provides a fixed hyperplane $H \subset \mathbb{R}^l \setminus (\mathbb{R}_+^l \setminus \{0\})$ such that all his convex upper contour sets G_t lie on one side of H, namely on the same side as $\mathbb{R}_+^l$. Moreover, every G_t is contained in a strictly convex cone L_t^* with vertex 0 which also lies on the same side of H as G_t and intersects H only in 0. Furthermore, the set $G_t \cup \partial L_t^*$ is a certain set of image points of the excess demand function under consideration. Actually, this ensures that the preference relation represented by the sets G_t generates the excess demand function.

For any t two mappings

$$\lambda_t : H \to \mathbb{R}_+$$

$$\gamma_t : H \to \mathbb{R}_+$$

are considered where $\lambda_t(q)$ is the least $s' \in \mathbb{R}_+$ such that $q + s'e \in L_t^*$ (recall $e = (1,1,\ldots,1)$) and $\gamma_t(q)$ is the least $s'' \in \mathbb{R}_+$ such that $q+s''e \in G_t$. λ_t and γ_t are convex functions, $\lambda_t \le \gamma_t$, and $\lambda_t(q) = \gamma_t(q)$ is equivalent to $q + \lambda_t(q)e \in G_t \cap \partial L_t^*$ (cf. Figure 23.5 below). Finally, a fixed continuous and convex function

$$\rho : \{(x,y) \in \mathbb{R}_+^2 | x \le y\} \to \mathbb{R}_+$$

is introduced which makes

$$\mu_t : H \to \mathbb{R}_+$$
$$(p,q) \mapsto \rho[\lambda_t(q), \gamma_t(q)]$$

continuous and strictly convex. Then

$$M_t := \{q + \bar{s}e | q \in H, \bar{s} \ge \mu_t(q)\}$$

is the new strictly convexified upper contour set.

Now we have to verify that this procedure is also applicable to our situation here: actually, the role of H is played now by the normal hyperplane $T_{\bar{p}^i}$ to $\bar{p}^i$. From the properties of v_i follows that for any $p^1 \in S_{++}^{l-1}$ the preimage $v_i^{-1}]-\infty, v_i(p^1)] = u_i^{-1}[0, u_i(p^1)] \subset S_{++}^{l-1}$ spans a strictly convex cone (with vertex 0) such that $grad\, v_i(p^1) \in T_{\bar{p}^1}$ is

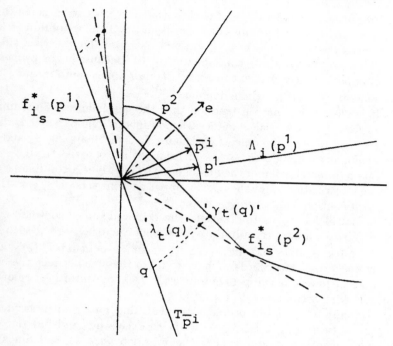

Fig. 23.5: Auxiliary Constructions II for Proving Lemma 23.3

normal to a supporting hyperplane S_{p^1} at p^1 of that cone (see Mas-Colell, 1977, p. 121). Figure 23.5 illustrates that for $l = 2$ (for this low dimension the cone $v_i^{-1}]-\infty, v_i(p^1)]$ actually is only convex).
The role of the upper contour sets G_t is played here by the upper contour sets contained in $\psi_{i_s}'''(0)$, i.e. by the upper contour sets $\psi_{i_s}'''(f_{i_s}^*(p))$, $p \in S_{++}^{l-1}$, and those which are 'better' than these.
The role of the sets L_t^* is played by the cones

$$\Lambda_i(p^1) := \{x \in \mathbb{R}^l | xp \geq 0 \text{ for all } p \in u_i^{-1}[0, u_i(p^1)]\},$$

$p^1 \in S_{++}^{l-1}$. By construction $\Lambda_i(p^1) = \Lambda_i(p^2)$ if and only if $u_i(p^1) = u_i(p^2)$. Furthermore one has

$$\psi_{i_s}'''(f_{i_s}^*(p^1)) \cap \partial\Lambda_i(p^1) = f_{i_s}^*(u_i^{-1}(u_i(p^1))).$$

(Clearly, $u_i^{-1}(u_i(p^1))$ is the boundary of $u_i^{-1}[0, u_i(p^1)] \subset S_{++}^{l-1}$.) Also by construction the set-valued function

$$p \mapsto u_i^{-1}[0, u_i(p)]$$

is Hausdorff continuous, and hence also $p^1 \mapsto \Lambda_i(p^1)$ is Hausdorff continuous.

From these considerations it is clear that Debreu's strict convexification process directly applies to our situation. Thus we obtain strictly convex upper contour sets $\psi''_{i_s}(x)$ which are on the 'positive' side of $T_{\bar{p}^i}$, and, moreover, are contained in $\psi''_{i_s}(0)$. Because of the properties of ρ, particularly that of convexity, the new upper contour sets $\psi''_{i_s}(x)$ generate $f^*_{i_s}$ on the whole domain S^{l-1}_{++}.

It is clear from the construction that the continuous one-parametrization $(\preceq'''_{i_s})_{s\in[0,1]}$ also induces a continuous one-parametrization $(\preceq''_{i_s})_{s\in[0,1]}$: the reference hyperplane $T_{\bar{p}^i}$, the mapping ρ, and the analogous mapping to λ_t are invariant when s varies in $[0,1]$. The one-parametrization of the analogue to γ_t clearly is a continuous one-parametrization since the convex upper contour sets of $\preceq'''_{i_s}$ are Hausdorff continuously varying with s.

(3) It is still left to us to translate the achieved upper contour sets of $\preceq''_{i_s}$ by an appropriate initial endowment vector $\omega_i \in \mathbb{R}^l_+$ such that for all $s \in [0,1]$ and all $p \in S^{l-1}_{++}$ agent i's demand $f^*_{i_s}(p) + \omega_i$ is nonnegative. Afterwards we have to fill in the hole between the origin and the lowest new upper contour set by Debreu's intuitive homothetic shrinking technique (1974, p. 20).

Actually, there is no problem at all to find an ω_i with the desired properties: any $f^*_{i_s} : S^{l-1}_{++} \to \mathbb{R}^l$ is an excess demand function, and consequently it is bounded from below, say by $k_{i_s}e$, $k_{i_s} < 0$. Since $[0,1]$ is compact, also the whole continuous one-parametrization $(f^*_{i_s})_{s\in[0,1]}$ is uniformly bounded from below, say by $k_i e$, $k_i < 0$. Providing agent i with the constant initial endowment l-vector ω_i with all components equal to $|k_i|$, for instance, pushes his individual demand

$$f^*_{i_s}(p) + \omega_i = f^*_{i_s}(p) + \begin{pmatrix} |k_i| \\ |k_i| \\ \vdots \\ |k_i| \end{pmatrix}$$

for all $p \in S^{l-1}_{++}$ and $s \in [0,1]$ into $\mathbb{R}^l_+$. Accordingly we translate all upper contour sets $\psi''_{i_s}(x)$ by ω_i, i.e. we replace $\psi''_{i_s}(x)$ by $\psi'_{i_s}(x) := \psi''_{i_s}(x) + \omega_i$. Clearly, $f^*_{i_s}$ is generated by the obtained pair $(\preceq'_{i_s}, \omega_i)$.

(4) Now, in a last step, we restrict the preference relation $\preceq'_{i_s}$ to $\mathbb{R}^l_+$ by intersecting its upper contour sets with $\mathbb{R}^l_+$, and fill in the hole between 0 and $\Omega_i := (\psi''_{i_s}(0) + \omega_i) \cap \mathbb{R}^l_+$ using Debreu's homothetic radial shrinking technique: we just take the strictly convex sets $t\Omega_i$, $t \in]0, ||\omega_i||]$. Thus, a preference relation $\preceq_{i_s}$ on $\mathbb{R}^l_+$ obtains such that $(\preceq_{i_s}, \omega_i)$ generates $f^*_{i_s}$ on S^{l-1}_{++}.

Let us summarize: for any $i \in \{1, \ldots, l\}$ and any $s \in [0,1]$ we have achieved a pair $(\preceq_{i_s}, \omega_i)$ with $\preceq_{i_s} \in \mathcal{P}^0_{\substack{mo \\ sco}}$ such that $(\preceq_{i_s}, \omega_i)$ generates $f^*_{i_s}$ on the whole price space S^{l-1}_{++}. Moreover, the one-parametrization

$(\preceq_{i_s})_{s\in[0,1]}$ is continuous, and thus we have obtained a continuous one-parametrization of exchange economies

$$(E_s)_{s\in[0,1]} = \left((\preceq_{i_s}, \omega_{i_s})_{i=1}^l \right)_{s\in[0,1]}$$

$$= \left(\left(\preceq_{i_s}, \begin{pmatrix} |k_i| \\ \vdots \\ |k_i| \end{pmatrix} \right)^l_{i=1} \right)_{s\in[0,1]}$$

with the desired properties.

$\square$

Proof of Corollary 13.2

Pulling back the unbounded price space $\mathbb{R}_+^l \setminus \{0^l\}$ to $T_0^l \setminus \Delta^{l-1} \subset \mathbb{R}_+^l$ by means of the homeomorphism φ^{-1}, and replacing T^l by S_+^l, Part (I) of the proof of Theorem 13.1 can immediately be adapted to the present situation.

$\square$

Proof of Proposition 13.4

Let us denote the price space by P and the two given compact subsets by K_1 and K_2. Now consider the convex hull M of K_1 and K_2 in P, and the compact subspace

$$K := K_1 \times [0, 1/2[\ \cup \ M \times \{1/2\} \ \cup \ K_2 \times]1/2, 1]$$

of P. Clearly K qualifies for Theorem 13.1 and Corollary 13.2, respectively, and Proposition 13.4 is proved.

$\square$

Appendix C

Proof of Proposition 14.3

We have to show that for any regular path $\eta = (\eta_s)_{s \in [0,1]} \in J'$ there is an open neighborhood V_η in J' so that ψ is constant on V_η, i.e. ψ is locally constant. Consider first the two bordering regular economies η_0 and η_1 of η. Denote their equilibrium sets by E_{η_0} and E_{η_1} respectively. Clearly, there are real numbers $\epsilon_0, \epsilon_1 > 0$ such that η_s remains in the same regular component of the space of economies as η_0 which implies $\#E_{\eta_0} = \#E_{\eta_s}$ for $0 \leq s \leq \epsilon_0$ and correspondingly $\#E_{\eta_1} = \#E_{\eta_s}$ for $1 - \epsilon_1 \leq s \leq 1$ (see Figure 24.1). Furthermore, $\#E_{\eta_0}$ and $\#E_{\eta_1}$ are odd.

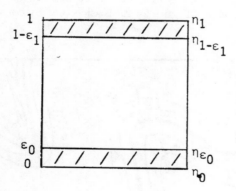

Fig. 24.1: Auxiliary Constructions I for Proving Proposition 14.3

Now let us look at the equilibrium set of the regular path η. We know that up to diffeomorphisms it consists of finitely many disjoint joining segments, running back segments, and circles such that only the joining segments and the running back segments intersect the border faces $S^{l-1} \times \{0\}$ and $S^{l-1} \times \{1\}$ of the homotopy space.

We are only interested in changes of the number of joining segments. Clearly, such changes must be either due to *vanishing* or to *appearing joining segments*. The key for our subsequent argumentation is provided by the following straightforward observation:

(+) *from the oddness of the number of equilibria of a regular economy follows directly that if η is changed in J' so carefully that the bordering economies are not pushed out of their respective regular components, then joining segments can only appear pairwise so that one running back segment simultaneously vanishes on each border face of the homotopy space. Correspondingly, joining segments can only vanish pairwise so that one running back segment simultaneously appears on each side of the homotopy space.*

After these preparatory observations we begin with the main line of our proof. Actually, the whole proof bases on the simple idea that vanishing or appearing joining equilibrium segments must generate points in areas which, nevertheless, can be shown to remain free of equilibria for paths $\bar\eta$ close to η. Denote the equilibrium set of η by K and its components by $k, j, \ldots$. Let be $\delta := \min[d[K, \partial S^{l-1}_+ \times [0,1]], \min_{k,j \in K} d(k,j)]$. Since K is a compact subset of $S^{l-1}_{++} \times [0,1]$ and any two $k, j \in K$ are compact and disjoint we know that $\delta > 0$. Now, for any component $k \in K$ we consider its closed relative $1/5$-*tubular neighborhood* $D^1_k := \{(x,s) \in S^{l-1}_+ \times [0,1] | d(k,(x,s)) \leq 1/5\delta\}$ and its closed relative $2/5\delta$-tubular neighborhood D^2_k. Consider furthermore the compact difference set $Z_k := D^2_k \setminus \overset{\circ}{\Delta}{}^1_k \subset S^{l-1}_+ \times [0,1]$ which we will call the *compact tubular jacket* of k (see the shaded areas in Figure 24.2). Notice that by construction for any k

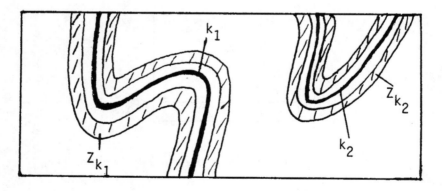

Fig. 24.2: Auxiliary Constructions II for Proving Proposition 14.3

we have $Z_k \cap K = \emptyset$, and $Z_k \cap Z_j = \emptyset$ for any two components $k \neq j$ from K.

Now pick any $k \in K$ and any point $(\overline{x}, \overline{s})$ from Z_k. Since $Z_k \cap K = \emptyset$ there must be at least one market excess demand function $\zeta_{\overline{s}_i}$ of $(\eta_s)_{s \in [0,1]}$ for some commodity i with

$$\zeta_{\overline{s}_i}(\overline{x}) > 0, \quad \text{or} \quad \zeta_{\overline{s}_i}(\overline{x}) < 0.$$

Without loss of generality let us assume that $\zeta_{\overline{s}_i}(\overline{x}) > 0$. We also say that the commodity number i is *associated with* $(\overline{x}, \overline{s})$. Since $\zeta_{\overline{s}_i}(-)$ is *continuous* there is an open neighborhood $\mathcal{U}(\overline{x}, \overline{s})$ of $\zeta_{\overline{s}_i}$ in the space of market excess demand functions such that, furthermore, there is clearly a relative open neighborhood $\mathcal{V}(\overline{x}, \overline{s})$ of $\overline{x}$ in S_{++}^{l-1} such that

$$\forall_{\substack{f_i \in \mathcal{U}(\overline{x}, \overline{s}) \\ y \in \mathcal{V}(\overline{x}, \overline{s})}} f_i(y) > 0.$$

Since $\bigcup_{k \in K} Z_k \subset S_+^{l-1} \times [0,1]$ is compact, a finite number of such open neighborhoods $\mathcal{V}(x_1, s_1), \mathcal{U}(x_1, s_1), \ldots, \mathcal{V}(x_r, s_r), \mathcal{U}(x_r, s_r)$ suffices "to cover" $\bigcup_{k \in K} Z_k$, i.e. $\bigcup_{k \in K} Z_k \subseteq \bigcup_{h=1}^{r} (\mathcal{V}(x_h, s_h) \times I_h)$ with $I_h := \{s \in [0,1] | \eta_{s_i} \in \mathcal{U}(x_h, s_h)\} \subseteq [0,1]$. (Clearly, all intervals I_h are non-empty and $\bigcup_{h=1}^{r} I_h = [0,1]$.)

Let us now choose two open neighborhoods $\mathcal{U}_0$, $\mathcal{U}_1$ of the bordering economies η_0 and η_1 in the space of economies which are contained in the respective regular components of η_0 and η_1 with the property that no equilibrium leaves its relative open $1/5\delta$-neighborhood in S_{++}^{l-1} when η_0 (η_1) is varied in $\mathcal{U}_0$ ($\mathcal{U}_1$). Choose furthermore two open neighborhoods $\widetilde{\mathcal{U}}_0$, $\widetilde{\mathcal{U}}_1$ of the path η in J' such that $\widetilde{\mathcal{U}}_0|_{\eta_0} \subseteq \mathcal{U}_0$ and $\widetilde{\mathcal{U}}_1|_{\eta_1} \subseteq \mathcal{U}_1$.

Now choose for any $h = 1, \ldots, r$ an open neighborhood W_h of the given regular path η in J' whose restriction to the s_h-state economy and to that (market) component function whose index is associated with (x_h, s_h) equals $\mathcal{U}(x_h, s_h)$.

Now consider the open neighborhood

$$V_\eta := \bigcap_{h=1,\ldots,r} W_h \cap \widetilde{\mathcal{U}}_0 \cap \widetilde{\mathcal{U}}_1$$

of η in J'. By construction clearly

$$(*) \qquad \forall_{\overline{\eta} \in V_\eta} \ Equ_{\overline{\eta}} \cap \left(\bigcup_{k \in K} Z_k \right) = \emptyset$$

where $Equ_{\overline{\eta}}$ denotes the equilibrium set of $\overline{\eta}$. But this means that for any $\overline{\eta}$ from the neighborhood V_η of η the number of joining equilibrium segments remains constant. Otherwise property $(*)$ would be violated due to our observation $(+)$ from above. This completes our proof.

$\square$

References

1. Allen B (1981) Utility Perturbations and the Equilibrium Price Set. J. of Math. Econ. 8:277–307
2. Allgower E, Georg K (1980) Simplicial and Continuation Methods for Approximating Fixed Points and Solutions to Systems of Equations. SIAM Review 22, 1:28–85
3. Arrow KJ, Hahn F (1971) Competitive Equilibrium Analysis. Holden-Day
4. Balasko Y (1975a) The Graph of the Walras Correspondence. Econometrica 43:907–912
5. Balasko Y (1975b) Some Results on Uniqueness and on Stability of Equilibrium in General Equilibrium Theory. J. of Math. Econ. 2:95–118
6. Balasko Y (1978a) Economic Equilibrium and Catastrophe Theory: An Introduction. Econometrica 46:557–569
7. Balasko Y (1978b) The Behaviour of Economic Equilibria: A Catastrophe Theory Approach. Behavioral Science 23:375–382
8. Balasko Y (1978c) Equilibrium Analysis and Envelope Theory. J. of Math. Econ. 5:153–172
9. Balasko Y (1979) A Geometric Approach to Equilibrium Analysis. J. of Math. Econ. 6:217–228
10. Balasko Y (1980) Number and Definiteness of Economic Equilibria. J. of Math. Econ. 7:215–225
11. Balasko Y (1988) Foundations of the Theory of General Equilibrium. Academic Press
12. Balasko Y (1996) Equilibres et Discontinuité. Campus. Magazine de l'Université de Genève, 31:46–47
13. Balasko Y, Lang Chr (1998) Manifolds of Golden Rule and Balanced Steady State Equilibria. Econ. Theory 11:317–330
14. Batten D, Casti J, Johansson B. (eds) (1987) Economic Evolution and Structural Adjustment. Proceedings, Lecture Notes in Economics and Mathematical Systems vol. 293, Springer-Verlag
15. Baumol W (1970) Economic Dynamics. 3. ed., Macmillan
16. Benassy JP (1982) The Economics of Market Disequilibrium. Series Economic Theory, Econometrics, and Mathematical Economics, Academic Press
17. Böhm V (1982) On the Uniqueness of Macroeconomic Equilibria with Quantity Rationing. Economics Letters 10:43–48

18. Böhm V (1989) Disequilibrium and Macroeconomics. Basil Blackwell
19. Bonnisseau JM, Cayupi JR (1999) The Equilibrium Manifold with Boundary Constraints on the Consumption Sets. Cahiers Eco and Maths 97.92, Univ. de Paris I
20. Bosch A (1990) Market Process as an Evolutionary Process. in: General Equilibrium or Market Process: Neoclassical and Austrian Theories of Economics, ed. A. Bosch, Mohr Verlag, pp. 77–98
21. Browder F (1960) On Continuity of Fixed Points under Deformation of Continuous Mappings. Summa Brasiliensis Mathematicae 4:183–191
22. Brown RF (1971) The Lefschetz Fixed Point Theorem. Scott, Foresman, and Company; London
23. Christiansen DS, Majumdar MK (1977) On Shifting Temporary Equilibrium. Journ. of Econ. Th. 16:1–9
24. Day R (1987) The Evolving Economy. European Journal of Operational Research, 30:251–257
25. Debreu G (1959) Theory of Value. Wiley and Sons
26. Debreu G (1969) Neighboring Economic Agents. La Décision, C.N.R.S., Paris:85–90
27. Debreu G (1970) Economies with a Finite Set of Equilibria. Econometrica 38:387–392
28. Debreu G (1974) Four Aspects of the Mathematical Theory of Economic Equilibrium. Proceedings of the International Congress of Mathematicians, Vancouver 1974:65–77
29. Debreu G (1976) The Application to Economics of Differential Topology and Global Analysis: Regular Differentiable Economies. Amer. Econ. Reviews 66, Papers and Proceedings:280–287
30. Debreu G (1982) Existence of Competitive Equilibrium. Ch. 15 in Handbook of Mathematical Economics, Vol. II, edt. by Arrow and Intriligator, North-Holland
31. Debreu G (1984) Economic Theory in the Mathematical Mode. Nobel-Prize-Lecture, delivered in Stockholm on December 8, 1983, The Amer. Econ. Review 74:267–278
32. Dierker E (1972) Two Remarks on the Number of Equilibria of an Economy. Econometrica 40:951–953
33. Dierker E (1974) Topological Methods in Walrasian Economies. Lecture Notes in Economics and Mathematical Systems 92, Springer-Verlag
34. Dierker E (1976) Regular Economies. in: Frontiers in Quantitative Economics, edt. by M. Intriligator, North-Holland
35. Dierker E (1982) Regular Economies. Ch. 17 in Handbook of Mathematical Economics, Vol. II
36. Dierker E, Dierker H (1972) The Local Uniqueness of Equilibria. Econometrica 40:867–881
37. Eaves BC, Scarf HE (1976) The Solution of Systems of Piecewise Linear Equations. Mathematics of Operations Research 1:1–27
38. Faber M, Proops J (1998) Evolution, Time, Production, and the Environment. Springer-Verlag
39. Fellner W et alii (1981) Shock therapy or gradualism? A comparative approach to anti-inflational policies. Group of Thirty, New York, Occ. Paper No. 8
40. Fisher FM (1983) Disequilibrium Foundations of Equilibrium Economics. Econometric Society Monographs No. 6, Cambridge University Press

41. Fort MK (1950) Essential and Non Essential Fixed Points. American J. of Math. 72:315–322
42. Frisch R (1935) On the Notion of Equilibrium and Disequilibrium. Rev. of Econ. Studies 9:100–105
43. Fuchs-Seliger S (1983) On Continuous Utility Functions Derived from Demand Functions. J. of Math. Economics 12:19–32
44. Gandolfo G (1980) Economic Dynamics: Methods and Models. 2. edition, North-Holland
45. Gandolfo G, Padoan PC (1984) A Disequilibrium Model of Real and Financial Accumulation in an Open Economy. Lecture Notes vol. 236, Springer-Verlag
46. Gandolfo G, Petit ML (1988) The Optimal Degree of Wage-Indexation in the Italian Economy: Rerunning History by Dynamic Optimization. In: Growth Cycles and Multisectoral Economics: the Goodwin tradition, edt. by Ricci and Velupillai, pp 120–126, Lecture Notes vol. 309, Springer-Verlag
47. Ginsburgh V, Mercenier J (1987) Macroeconomic Models and Microeconomic Theory. Contributions of General Equilibrium Theory. Mimeographed manuscript, Universite Libre de Bruxelles
48. Goodwin RM (1948) Secular and Cyclical Aspects of the Multiplier and the Accelerator. in: Income, Employment, and Public Policy: Essays in Honor of A. Hansen, pp 108–132, Norton
49. Grandmont JM (1982) Temporary General Equilibrium Theory. Chapter 19 in Handbook of Mathematical Economics, vol. II
50. Hahn F (1982) Stability. Ch. 16 in Handbook of Mathematical Economics, vol. II
51. Hatta T (1977) A Theory of Piecemeal Policy Recommendations. Rev. of Ec. Studies 44:1–21
52. Hettich W (1979) A Theory of Partial Tax Reform. Canadian J. of Economics 12:692–712
53. Hicks J (1946) Value and Capital. Clarendon Press
54. Hicks J (1965) Capital and Growth. Clarendon Press
55. Hicks J (1984) The Economics of John Hicks. selected by D. Helm, Basil Blackwell
56. Hicks J (1985) Methods of Dynamic Economics. Clarendon Press
57. Hildenbrand W (1974) Core and Equilibria of a Large Economy. Princeton University Press
58. Hildenbrand W (1989) Facts and Ideas in Microeconomic Theory. European Economic Review 33:251–276
59. Hildenbrand W (1994) Market Demand: Theory and Empirical Evidence. Princeton University Press
60. Hildenbrand W (1998) How Relevant are Specifications of Behavioral Relations on the Micro-Level for Modelling the Time Path of Population Aggregates? European Economic Review 42:437–458
61. Hildenbrand W (1999a) An Introduction to Demand Aggregation. Journal of Mathematical Economics, Special Issue on Aggregation, 31:1–14
62. Hildenbrand W (1999b) On the Empirical Content of Economic Theories. In: Economics Beyond the Millennium, 37–54, Alan P. Kirman, L.-A. Gérard-Varet (eds.), Oxford University Press)
63. Hildenbrand W, Kirman A (1988) Equilibrium Analysis. North-Holland
64. Hirsch MW (1976) Differential Topology. Graduate Texts in Mathematics 33, Springer-Verlag

65. Jerison M (1984) Aggregation and Pairwise Aggregation when the Distribution of Income is Fixed. J. of Econ. Theory 33:1–31
66. Kehoe TJ (1980) An Index Theorem for General Equilibrium Models with Production. Econometrica 48:1211–1232
67. Kehoe TJ (1982) Regular Production Economies. J. of Math. Econ. 10:147–176
68. Kehoe TJ (1985a) Multiplicity of Equilibria and Comparative Statics. The Quarterly J. of Economics 100:119–148
69. Kehoe TJ (1985b) The Comparative Statics Properties of Tax Models. Canadian J. of Economics 18:314–334
70. Kirman A, Koch KJ (1986) Market Excess Demand in Exchange Economies with Identical Preferences and Collinear Endowments. Rev. of Econ. Studies 53:457–463
71. Kirzner I (1990) The Meaning of Market Process. In: General Equilibrium or Market Process: Neoclassical and Austrian Theories of Economics, ed. A. Bosch, Mohr Verlag, pp. 61–76
72. Kloek T (1984) Dynamic Adjustment when the Target is Nonstationary. Int. Ec. Rev. 25:315–326
73. Koch KJ (1989) Mean Demand when Consumers Satisfy the Weak Axiom of Revealed Preference. Journal of Math. Economics, 18:347–356
74. Kornai J (1983) Equilibrium as a Category of Economics. Acta Oeconomica 30:145–159
75. Lang S (1969) Analysis I. Addison-Wesley
76. Legendre F (1987) Dynamic Adjustment when the Target is Nonstationary: a Comment. Int. Ec. Rev. 28:809–811
77. Lehmann-Waffenschmidt M (1983) Fasernweise algebraische Fixpunktinvarianten und eine Anwendung in stetig deformierten Walrasianischen Ökonomien.. in: Mathematische Systeme in der Ökonomie, Athenäum Verlag:383–413, edt. by M. Beckmann, W. Eichhorn, W. Krelle
78. Lehmann-Waffenschmidt M (1985) Gleichgewichtspfade für Ökonomien mit variierenden Daten. series: mathematical systems in economics, vol. 99, Hain-Verlag
79. Lehmann-Waffenschmidt M (1987) Bounding the Price Space $\mathbb{R}_+^n$ by a Collar-Preserving Homeomorphism. in: Ökonomie und Mathematik, 49–62, edt. by Opitz, Rauhut; Springer-Verlag
80. Lehmann-Waffenschmidt M (1994) Existence and Stability of Equilibrium in a Multi-Sectoral Model with Quantity Rationing. in: Models and Measurement of Welfare and Inequality. ed. W. Eichhorn, Springer Verlag
81. Lehmann-Waffenschmidt M (1995) On the Equilibrium Price Set of Continuous Perturbations of Exchange Economies. Journal of Mathematical Economics, 24, pp. 497–519
82. Lehmann-Waffenschmidt M (2005) A Fresh Look on Economic Evolution from the Kinetic Viewpoint. Journal of Evolutionary Economics, vol. 15, 5, pp. 481–503
83. Lehmann-Waffenschmidt M (2006) A Note on Continuously Decomposed Evolving Exchange Economies. Central European Journal of Operations Research, 3, pp. 289–298
84. Leininger W (1978) Ein konstruktives Beweisprinzip der nichtlinearen Funktionalanalysis mit Homotopien. diploma thesis, Mathematical Department of the University of Bonn

85. Loasby BJ (1991) Equilibrium and Evolution. An exploration of connecting principles in economics Manchester University Press

86. Lojasiewicz S (1970) Sur les ensembles semi-analytiques. Actes Congrès intern. Math., Tome 2:237–241 (Section 8)

87. Lojasiewicz S (1971) Sur les ensembles semi-analytiques. Pubbl. dell'Istituto Nazionale di Alta Matematica, 1970 Symp. Math. Vol. III, 233–239 (Sect. 8)

88. Los J, Los MW (eds.) (1976) Computing Equilibria: How and Why. Proceedings of the International Conference in Torun, Poland, 8.–13.7.1974; North-Holland

89. Manne AS (ed.) (1985) Economic Equilibrium. Model Formulation and Solution. Mathematical Programming Studies, vol. 23, North-Holland

90. Mantel R (1974) On the Characterization of Aggregate Excess Demand. Journal of Econ. Theory 7:348–359

91. Mantel R (1976) Homothetic Preferences and Community Excess Demand Fucntions. J. of Econ. Theory 12:197–201

92. Marangos J (2002) The Political Economy of Shock Therapy. J. of Econ. Surveys, 16:41–76

93. Mas-Colell A (1974) A Note on a Theorem of F. Browder. Mathematical Programming 6:229–233

94. Mas-Colell A (1977) On the Equilibrium Price Set of an Exchange Economy. J. of Math. Econ. 4:117–126

95. Mas-Colell A (1985) The Theory of General Economic Equilibrium. A Differentiable Approach. Econometric Society Publication No. 9, Cambridge University Press

96. McFadden D, Mas-Colell A, Mantel R, Richter R (1974) A Characterization of Community Excess Demand Functions. J. of Econ. Theory 7:348–353

97. McKenzie LW (1981) The Classical Theorem on Existence of Competitive Equilibrium. Econometrica 49:819–841

98. O'Neill B (1953) Essential Sets and Fixed Points. Amer. J. of Math. 75:497–509

99. Ott AE (1970) Einführung in die dynamische Wirtschaftstheorie. 2. erw. Auflage, Vandenhoeck und Ruprecht

100. Puppe D (1979) Duality in Monoidal Categories and Applications in Homotopy Theory. in: Game Theory and Related Topics, edt. by Moeschlin and Pallaschke, North-Holland:173–185

101. Quirk J, Saposnik R (1968) Introduction to General Equilibrium Theory and Welfare Economics. Mc Graw-Hill

102. Rheinboldt WC (1986a) Numerical Analysis of Parametrized Nonlinear Equations. Wiley and Sons

103. Rheinboldt WC (1986b) On the Computation of Multi-dimensional Solution Manifolds of Parametrized Equations. Num. Math. (also available as Tech. Report ICMA-86-102, 1986, Inst. for Comput. Math. and Applic., Univ. of Pittsburgh)

104. Rheinboldt WC (1987) Error Questions in the Computation of Solution Manifolds of Parametrized Equations. in: Reliability in Computing ed. by R.A. Moore, Academic Press (also available as Tech. Report ICMA-87-112, 1987, Inst. for Comput. Math. and Applic., Univ. of Pittsburgh)

105. Rosenstein-Rodan PN (1930) Das Zeitmoment in der Mathematischen Theorie des wirtschaftlichen Gleichgewichts. Zeitschrift für Nationalökonomie 1:129–142

106. Samuelson PA (1941) The Stability of Equilibrium: Comparative Statics and Dynamics. Econometrica 9:97–120
107. Samuelson PA (1947) Foundations of Economic Analysis. Cambridge Harvard University Press
108. Samuelson PA (1948) Dynamic Process Analysis. in: A Survey of Contemporary Economics, edt. by H.S. Ellis, The Blakiston Company
109. Scarf H (1960) Some Examples of Global Instability of the Competitive Equilibria. Intern. Ec. Rev. 1:157–172
110. Scarf H (1982) The Computation of Equilibrium Prices: An Exposition. Ch. 21 in Handbook of Mathematical Economics, vol. II, North-Holland
111. Schecter S (1979) On the Structure of the Equilibrium Manifold. J. of Math. Econ. 6:1–5
112. Schirmer H (1983) Fixed Point Sets of Homotopies. Pacific J. of Mathematics 108:191–202
113. Schulz N (1982) Market Demand and Excess Demand Functions. Chapter 14 in Handbook of Mathematical Economics, vol. II
114. Schulz N (1983) On the Global Uniqueness of Fix-Price Equilibria. Econometrica 51:47–68
115. Schulz N (1985) Existence of Equilibria Based on Continuity and Boundary Behaviour. Economics Letters 19:101–103
116. Shafer W, Sonnenschein H (1976) Equilibrium with externalities, commodity taxation, and lump sum transfers. Inter. Economic Review 17:601–611
117. Shoven JB, Whalley J (1973) General Equilibrium with Taxes: A Computational Procedure and an Existence Proof. Rev. of Ec. Studies 40:475–489
118. Shoven JB, Whalley J (1984) Applied General-Equilibrium Models of Taxation and International Trade: An Introduction and Survey. J. of Econ. Lit. 22, pp. 1007–1051
119. Smale S (1976) A Convergent Process of Price Adjustment and Global Newton Methods. Journ. of Math. Economics 3:107–120
120. Smale S (1980) The Mathematics of Time. Springer-Verlag
121. Smale S (1981) Global Analysis and Economics. Ch. 8 in Handbook of Mathematical Economics, vol. I, edt. by Arrow and Intriligator, North-Holland
122. Sonnenschein H (1973a) The Utility Hypothesis and Market Demand Theory. Western Econ. Journal, 11:404–410
123. Sonnenschein H (1973b) Do Walras' Identity and Continuity Characterize Community Excess Demand Functions?. J. of Econ. Theory 6:345–354
124. Sonnenschein H (1974) Market Excess Demand Functions. Econometrica 40:549–563
125. Sonnenschein H (ed.) (1986) Models of Economic Dynamics. Proceedings, Minneapolis, USA, October 1983, Lecture Notes in Economics and Mathematical Systems, Vol. 254, Springer-Verlag
126. Talman AJJ, Laan Gvd (eds.) (1987) The Computation and Modelling of Economic Equilibria. North-Holland
127. Teissier B (1975) Théorèmes des finitude en géometrie analytique (d'après Heisuke Hironaka). in: Séminaire Bourbaki, vol. 1973/74, Exposées 436–452; Lecture Notes in Mathematics, vol. 431, Springer-Verlag
128. Todd MJ (1976) The Computation of Fixed Points and Applications. Lecture Notes in Economics and Mathematical Systems, vol. 124, Springer-Verlag
129. Todd MJ (1979) A Note on Computing Equilibria in Economies with Activity Analysis Models of Production. Journ. of Math. Econ. 6, 135–144

130. Uzawa H (1962) Walras' Existence Theorem and Brouwer's Fixed-Point Theorem. Econ. Studies Quarterly 13:59–62
131. Varian, H.R. (1973) A Third Remark on the Number of Equilibria of an Economy. Econometrica 43:985–986
132. Walras L (1954) 1874–1877. Elements of Pure Economics. translated by W. Jaffé, Homewood, R. Irwin
133. Weintraub R, Gayer T (2001) Equilibrium Proofmaking. J. of History of Econ. Thought, 23:421–442
134. Witt U (1992) Evolutionary Concepts in Economics. Eastern Economic Journal, Vol. 18:405-419
135. Witt U (2003) The Evolving Economy. Edward Elgar, Cheltenham
136. Zodrow GR (1985) Partial Tax Reform: an Optimal Taxation Perspective. Canadian J. of Economics 18:335–346

Subject Index

This index compiles the key terms of the book. The references indicate pages where the respective terms play a central role. The index is not supposed to give an exhaustive list of all terms and places in the book where a term is used. Particularly, not all terms defined in the Section "Mathematical Preliminaries" are listed in this Subject Index.

Lecture Notes in Economics and Mathematical Systems

For information about Vols. 1–502
please contact your bookseller or Springer-Verlag